Algebraic Methods in Cryptography

CONTEMPORARY MATHEMATICS

418

Algebraic Methods in Cryptography

AMS/DMV Joint International Meeting
June 16–19, 2005
Mainz, Germany

International Workshop on Algebraic Methods
in Cryptography
November 17–18, 2005
Bochum, Germany

Lothar Gerritzen
Dorian Goldfeld
Martin Kreuzer
Gerhard Rosenberger
Vladimir Shpilrain
Editors

American Mathematical Society
Providence, Rhode Island

Editorial Board

Dennis DeTurck, managing editor

George Andrews Carlos Berenstein Andreas Blass Abel Klein

2000 *Mathematics Subject Classification.* Primary 94A60, 20Fxx, 68P25, 68W20, 68W30, 11T71, 57M05.

Library of Congress Cataloging-in-Publication Data

Special Session on Algebraic Cryptography at the Joint International Meeting of the AMS and the Deutsche Mathematiker-Vereinigung (2005 : Mainz, Germany)
  Algebraic methods in cryptography : Special Session on Algebraic Cryptography at the Joint International Meeting of the AMS and the Deutsche Mathematiker-Vereinigung, June 16–19, 2005, Mainz, Germany : International Workshop on Algebraic Methods in Cryptography, November 17–18, 2005, Bochum, Germany / editors, Lothar Gerritzen ... [et al.].
    p. cm. — (Contemporary mathematics, ISSN 0271-4132 ; v. 418)
  ISBN-13: 978-0-8218-4037-5 (alk. paper)
  ISBN-10: 0-8218-4037-1 (alk. paper)
  1. Algebra. 2. Cryptography. I. Gerritzen, Lothar, 1941– II. American Mathematical Society. III. Deutsche Mathematiker-Vereinigung. IV. International Workshop on Algebraic Methods in Cryptography (2005 : Bochum, Germany) V. Title.

QA150.S64 2005
652′.8—dc22   2006043028

Copying and reprinting. Material in this book may be reproduced by any means for educational and scientific purposes without fee or permission with the exception of reproduction by services that collect fees for delivery of documents and provided that the customary acknowledgment of the source is given. This consent does not extend to other kinds of copying for general distribution, for advertising or promotional purposes, or for resale. Requests for permission for commercial use of material should be addressed to the Acquisitions Department, American Mathematical Society, 201 Charles Street, Providence, Rhode Island 02904-2294, USA. Requests can also be made by e-mail to reprint-permission@ams.org.

Excluded from these provisions is material in articles for which the author holds copyright. In such cases, requests for permission to use or reprint should be addressed directly to the author(s). (Copyright ownership is indicated in the notice in the lower right-hand corner of the first page of each article.)

© 2006 by the American Mathematical Society. All rights reserved.
The American Mathematical Society retains all rights
except those granted to the United States Government.
Copyright of individual articles may revert to the public domain 28 years
after publication. Contact the AMS for copyright status of individual articles.
Printed in the United States of America.

∞ The paper used in this book is acid-free and falls within the guidelines
established to ensure permanence and durability.
Visit the AMS home page at http://www.ams.org/

10 9 8 7 6 5 4 3 2 1   11 10 09 08 07 06

Contents

Key agreement, the Algebraic Eraser™, and lightweight cryptography
    IRIS ANSHEL, MICHAEL ANSHEL, DORIAN GOLDFELD,
    STEPHANE LEMIEUX   1

Designing Key Transport Protocols Using Combinatorial Group Theory
    G. BAUMSLAG, T. CAMPS, B. FINE, G. ROSENBERGER, AND X. XU   35

Geometric Key Establishment
    ARKADY BERENSTEIN AND LEON CHERNYAK   45

Using shifted conjugacy in braid-based cryptography
    PATRICK DEHORNOY   65

Length-based conjugacy search in the braid group
    DAVID GARBER, SHMUEL KAPLAN, MINA TEICHER, BOAZ TSABAN, AND
    UZI VISHNE   75

Towards Provable Security for Cryptographic Constructions Arising from Combinatorial Group Theory
    MARÍA ISABEL GONZÁLEZ VASCO, RAINER STEINWANDT, AND JORGE L.
    VILLAR   89

Constructions in public-key cryptography over matrix groups
    DIMA GRIGORIEV AND ILIA PONOMARENKO   103

A Practical Attack on the Root Problem in Braid Groups
    ANJA GROCH, DENNIS HOFHEINZ, AND RAINER STEINWANDT   121

An attack on a group-based cryptographic scheme
    DENNIS HOFHEINZ AND DOMINIQUE UNRUH   133

Algebraic Problems in Symmetric Cryptography: Two Recent Results on Highly Nonlinear Functions
    NILS GREGOR LEANDER   141

Inverting the Burau and Lawrence-Krammer Representations
    EONKYUNG LEE   153

A new key exchange protocol based on the decomposition problem
    VLADIMIR SHPILRAIN AND ALEXANDER USHAKOV   161

Using the subgroup membership search problem in public key cryptography
    VLADIMIR SHPILRAIN AND GABRIEL ZAPATA   169

Preface

This volume consists of contributions by speakers at the Special Session on Algebraic Cryptography at the Joint International Meeting of the AMS with the Deutsche Mathematiker-Vereinigung held in Mainz, Germany, on June 16–19, 2005, and at the International Workshop on Algebraic Methods in Cryptography held in Bochum, Germany, on November 17–18, 2005.

The readers will find here a variety of contributions, mostly related to public-key cryptography, including design of new cryptographic primitives as well as cryptanalysis of previously suggested schemes. Most papers are original research papers in the area that can be loosely defined as "Non-commutative cryptography"; this means that groups (or other algebraic structures) which are used as platforms are non-commutative.

We are grateful to the American Mathematical Society for assisting us in publication of this volume. In particular, we thank Christine M. Thivierge for her patient work in putting this volume together.

Lothar Gerritzen
Dorian Goldfeld
Martin Kreuzer
Gerhard Rosenberger
Vladimir Shpilrain

KEY AGREEMENT, THE ALGEBRAIC ERASER^{TM}, AND LIGHTWEIGHT CRYPTOGRAPHY

Iris Anshel, Michael Anshel, Dorian Goldfeld, Stephane Lemieux

§1. Introduction:

Our purpose is to present a new key agreement protocol for public-key cryptography suitable for implementation on low-cost platforms which constrain the use of computational resources. In the process we introduce the concept of an Algebraic Eraser^{TM}, AE, and make a case that AE is a suitable primitive for use within lightweight cryptography. Our underlying motivation is the need to secure networks which deploy Radio Frequency Identification (RFID) tags used for identification, authentication, tracing and point-of-sale applications. The reader should consult [GJP] and [OSK] for further discussion.

The idea behind AE is to deny the cryptanalyst certain algebraic information inherent in many contemporary algebraic key agreement protocols employing group-theoretic transformations such as discrete exponention certain finite abelian groups or conjugation on certain infinite groups including braid groups (see [KM]). AE employs certain groups, homomorphisms, and actions of groups on monoids which to date force the cryptanalyst to primarily employ exhaustive search to determine the key. After careful formulation of the basic structure of AE we specify a general key agreement protocol based on the AE and go on to give some explicit examples including possible attacks and choice of secure parameters.

§2. The Algebraic Eraser^{TM} and its Associated Protocol:

The concept of the Algebraic Eraser emerges naturally when considering the following structures in tandem. Let M, N denote monoids and let S denote a group which acts on M on the left, and does not act on N. Given elements $s \in S$ and $m \in M$, we denote the result of s acting on m by $^s m$. The semidirect product of M and S, $M \rtimes S$ is defined to be the monoid whose underlying set is $M \times S$ and whose internal binary operation is given by

$$(m_1, s_1) \circ (m_2, s_2) = (m_1 {}^{s_1}m_2, s_1 s_2).$$

The authors would like to thank SecureRF for its support of this research. The authors would also like to thank Alan Silvester for doing a lot of the C++ coding.

The direct product of N and S is denoted by $N \times S$.

The *algebraic eraser* **E** is the binary operation specified within the 6-tuple,
$$(M \rtimes S, N, \Pi, \mathbf{E}, A, B),$$
termed the **E**-*structure*, where $M \rtimes S$ and N are as above, Π is (an easily computable) monoid homomorphism
$$\Pi : M \to N,$$
E is the function
$$\mathbf{E} : (N \times S) \times (M \rtimes S) \to N \times S$$
given by
$$\mathbf{E}\big((n,s), (m_1, s_1)\big) = \big(n\, \Pi({}^s m_1), s s_1\big),$$
and A, B are submonoids of $M \rtimes S$ such that for all $(a, s_a) \in A, (b, s_b) \in B$
$$(1) \qquad \mathbf{E}\big((\Pi(a), s_a), (b, s_b)\big) = \mathbf{E}\big((\Pi(b), s_b), (a, s_a)\big).$$

The submonoids A and B, which satisfy (1) above, are termed **E**-*Commuting*. For simplicity we will use the notation $\star$ as follows:
$$\mathbf{E}\big((n,s), (m_1, s_1)\big) = (n, s) \star (m_1, s_1).$$

The operation $\star$ satisfies the property that given $(n, s) \in N \times S$ and $(m_1, s_1), (m_2, s_2) \in M \rtimes S$ then
$$(2) \qquad \Big((n, s) \star (m_1, s_1)\Big) \star (m_2, s_2) = (n, s) \star \Big((m_1, s_1) \circ (m_2, s_2)\Big).$$

The identity (2) is easily verified and allows one to compute $\star$ iteratively provided an element $(m, s) \in M \rtimes S$ is expressed as a product of generators.

The term *algebraic eraser* is a fitting description of the function **E** in that given an elements of $N \times S$,
$$(n, s), \qquad \mathbf{E}\big((n,s), (m_1, s_1)\big)$$
the element (m_1, s_1) cannot generally be recovered since the action of the element s on m_1 is not visible once the function Π has been applied to ${}^s m_1$ i.e., the action of s on m_1 has been effectively erased.

With the algebraic eraser **E** and its associated **E**-*structure* specified we are in a position to introduce an associated key agreement protocol, **E**–KAP. Referring to the protocol users as Alice and Bob, each user is assigned a submonoid of N, N_A and N_B respectively so that N_A and N_B commute. Furthermore Alice and Bob are assigned the **E**–commuting submonoids A and B, respectively, which are determined by the fixed **E**–structure. With these assignments in place Alice and Bob choose their respective private keys which take the form
$$A_{\text{Private}} = \text{Alice's Private Key} = \Big(n_a, \langle (a_1, s_{a_1}), (a_2, s_{a_2}), \dots, (a_k, s_{a_k}) \rangle \Big)$$
and
$$B_{\text{Private}} = \text{Bob's Private Key} = \Big(n_b, \langle (b_1, s_{b_1}), (b_1, s_{b_2}), \dots, (b_\ell, s_{b_\ell}) \rangle \Big)$$
where $n_a \in N_A$, $n_b \in N_B$,
$$(a_1, s_{a_1}), (a_2, s_{a_2}), \dots, (a_k, s_{a_k}) \in A,$$

and
$$(b_1, s_{b_1}), (b_2, s_{b_2}), \ldots, (b_\ell, s_{b_\ell}) \in B.$$
Having made these choices, Alice and Bob can then announce their respective public keys:

A_{Public} = Alice's Public Key = $(\cdots((n_a, \text{id})\star(a_1, s_{a_1}))\star(a_2, s_{a_2}))\star\cdots)\star(a_k, s_{a_k}) \in N \times S,$

B_{Public} = Bob's Public Key = $(\cdots((n_b, \text{id})\star(b_1, s_{b_1}))\star(b_2, s_{b_2}))\star\cdots)\star(b_\ell, s_{b_\ell}) \in N \times S,$

where id denoted the identity element in S. With this done Alice and Bob are now each in a position to compute the shared secret:

(3)
$$(\cdots((n_a, \text{id}) \cdot B_{\text{Public}}\star(a_1, s_{a_1})) \star (a_2, s_{a_2})) \star \cdots) \star (a_k, s_{a_k}) =$$
$$(\cdots((n_b, \text{id}) \cdot A_{\text{Public}} \star (b_1, s_{b_1})) \star (b_2, s_{b_2})) \star \cdots) \star (b_\ell, s_{b_\ell}),$$

where $\cdot$ denoted multiplication in $N \times S$. The identity (3) holds because the submonoids A, B where chosen to **E**–commute, and the submonoids N_A, N_B themselves commute.

§3. Algebraic Constructions

The **E**–structure $(M \rtimes S, N, \Pi, \mathbf{E}, A, B)$ and its associate key agreement protocol lend themselves naturally to various natural algebraic/categorical constructions. Furthermore when we focus on the case of M being a group and S being a (sub)group of automorphisms of the group, a generalized commutator emerges from the **E**–commuting condition.

The direct product of two **E**–structures, $\mathbf{E}_1$ and $\mathbf{E}_2$ yield a third **E**–structure:

$$(M_1 \rtimes S_1, N_1, \Pi_1, \mathbf{E}_1, A_1, B_1) \times (M_2 \rtimes S_2, N_2, \Pi_2, \mathbf{E}_2, A_2, B_2) =$$
$$\Big((M_1 \times M_2) \rtimes (S_1 \times S_2), N_1 \times N_2, \Pi_1 \times \Pi_2, \mathbf{E}_1 \times \mathbf{E}_2, A_1 \times A_2, B_1 \times B_2\Big).$$

Given a submonoid $H \leq M$ which is S invariant, there is a natural sub–**E**–structure of $(M \rtimes S, N, \Pi, \mathbf{E}, A, B)$ to consider:

$$(H \rtimes S, N, \Pi \downarrow_H, \mathbf{E} \downarrow_{(N \times S) \times (H \rtimes S)}, A \cap H, B \cap H).$$

Finally the concept of a image of an **E**–structure can be approached by starting with a homomorphism $\Psi : N \to N_0$ and considering the **E**–structure

$$(M \rtimes S, N_0, \Psi \circ \Pi, \mathbf{E}_0, A, B),$$

where $\Psi \circ \Pi$, denotes the composite of Ψ and Π.

In the case M is actually a group and the homomorphism Π is surjective, then another possible image can be defined. In this case we know that $N \cong M/K$ where $K \trianglelefteq M$. If $L \trianglelefteq M$ is a subgroup which is invariant under S then S acts on the group M/L and we can form $M/L \rtimes S$. This allows us to define an image of

$$(M \rtimes S, M/K, (M \to M/K), \mathbf{E}, A, B),$$

to be

$$\Big((M/L \rtimes S), M/LK, (M/L \to M/LK), \mathbf{E}_L, (AL/L) \rtimes S, (BL/L) \rtimes S\Big)$$

where unspecified homomorphisms are simply the natural homomorphisms.

When we again restrict ourselves to the case of a group, G and we assume the group S is actually a group of automorphisms of G, $S \leq \mathrm{Aut}(G)$ then the hypothesis of **E**–commuting takes the following form. Elements in the subgroups A, B can be written as
$$(a, \alpha), \quad (b, \beta)$$
where $a, b \in G$ and $\alpha, \beta \in \mathrm{Aut}(G)$. The function Π can be assumed to take the form $G \to G/K$, and the identity (1) becomes
$$\big(a\ \alpha(b)\big)K, \alpha \circ \beta\big) = \big((b\ \beta(a))K, \beta \circ \alpha\big).$$

The identity emerging from the first component leads naturally to the following generalization of the classical commutator. Given elements $x, y \in G$, and $(a, \alpha), (b, \beta) \in \mathrm{Aut}(G)$ define
$$C(\alpha, \beta, x, y) = x\ y\beta(x^{-1})\alpha(y^{-1}).$$

Clearly when $\alpha, \beta = \mathrm{id}$ we are reduced to the classical definition. Some analogues of the various classical commutator identities generalize as follows (and are left to the reader to verify). Let
$$\Omega(\alpha, \beta, x, y) = \alpha(x)\ y\ \beta(x)^{-1},$$

then we have

Proposition 1. *With the notation as above, the following identities hold:*

- $C(\alpha, \beta, x, y)^{-1} = C(\beta^{-1}, \alpha^{-1}, \alpha(y), \beta(x))$
- $C(\alpha, \beta, xy, z) = \Omega\big(\mathrm{id}, \beta, x, C(\alpha, \beta, y, z)\big)\ C(\mathrm{id}, \mathrm{id}, \beta(x), \alpha(z))$
- $C(\alpha, \beta, x, yz) = C(\mathrm{id}, \mathrm{id}, x, y)\ \Omega\big(\mathrm{id}, \alpha, y, C(\alpha, \beta, x, z)\big)$
- (identity of Hall–Witt type, see [MKS])

$$y^{-1}C\big(\mathrm{id}, \alpha, C(\alpha, \alpha, y, \alpha(x^{-1})), \alpha(z^{-1})\big)y$$
$$\cdot z^{-1}C\big(\mathrm{id}, \alpha, C(\alpha, \alpha, z, \alpha(y^{-1})), \alpha^2(x^{-1})\big)z$$
$$\cdot \alpha(x^{-1})C\big(\mathrm{id}, \alpha, C(\alpha, \alpha^2, \alpha(x), z^{-1}), \alpha(y^{-1})\big)\alpha(x)$$

Before delving into the examples of our protocol we present a brief aside regarding a group theoretic authentication method. Recall the protocol introduced in [AAG1]: users Alice and Bob each choose private elements a, b in assigned subgroups A, B of a group G and in the end agree on the commutator $[a, b]$ known only to the users. In the course of the protocol Alice actually obtains the conjugate $b^{-1}ab$ and hence is in a position to compute the element
$$b^{-1}ab \cdot a \cdot [a, b] = b^{-1}a^2b = (b^{-1}ab)^2.$$

Assuming that extraction of roots, in particular square roots, is known to be a difficult problem, Alice can forward the element $b^{-1}a^2b$ to Bob who can then conjugate by the inverse of his private key b^{-1} to obtain the element a^2. Thus Alice has effectively transmitted the square of her private key a^2 to Bob over an open channel. With this done, any choice of a hash function $\mathcal{H}$ (i.e., a one–way collision–free function) generates an authentication protocol in the spirit of [D]:

(i) Alice chooses an element $r \in G$ and sends Bob the the element $c = \mathcal{H}(ra^2r^{-1})$,

(ii) Bob chooses a random bit d and sends d to Alice,

(iii) If $d = 0$ Alice sends the element r and Bob verifies that $c = \mathcal{H}(ra^2r^{-1})$,

(iv) If $d = 1$ Alice sends the conjugate $s = ra^2r^{-1}$ and Bob verifies that $c = \mathcal{H}(s^2)$.

§4. Examples of Key Agreement based on the Algebraic Eraser^{TM}:

Fix an integer $n \geq 7$ and a prime $p > n$. As an example of an algebraic eraser $\mathbf{E}$ whose associated key agreement protocol merits attention we begin by considering a subgroup
$$M \leq \mathrm{GL}(n, \mathbb{F}_p(t)),$$
where $t = (t_1, \ldots, t_n)$. Also, let $S = S_n$ be the symmetric group on n symbols. The group S acts on $\mathrm{GL}(n, \mathbb{F}_p(t))$ by permuting the variables $\{t_1, \ldots, t_n\}$, and we shall assume that the action of S maps M to itself. Furthermore we assume that the semidirect product $M \rtimes S$ is finitely generated by some set of elements,

(4) $$\{(x_1(t), s_1), \ldots, (x_\lambda(t), s_\lambda)\}.$$

In this example, the monoid N is chosen to be
$$N = \mathrm{GL}(n, \mathbb{F}_p).$$

In order to define the homomorphism Π, we fix n elements in $\mathbb{F}_p$,
$$\tau_1, \tau_2, \ldots, \tau_n \in \mathbb{F}_p,$$
and then evaluate
$$\Pi : M \to N$$
by setting
$$t_1 = \tau_1, \quad t_2 = \tau_2, \quad \ldots, \quad t_n = \tau_n.$$
A crucial assumption needs to be made at this point.

Assumption τ: Let $\tau = (\tau_1, \tau_2, \ldots, \tau_n)$. We assume that $x_i(\tau)$, $x_i(\tau)^{-1}$ are well defined for all $i = 1, 2, \ldots, n$.

There are, of course, many possible choices for the commuting submonoids N_A, N_B, which need to be chosen. One elementary choice for N_A and N_B is to first fix a matrix $m_0 \in GL(n, \mathbb{F}_p)$ of order $p^n - 1$. Then let $N_A = N_B$ be the submonoid of all matrices of the form

(5) $$\ell_1 m_0^{k_1} + \ell_2 m_0^{k_2} + \cdots + \ell_r m_0^{k_r},$$

where $\ell_1, \ell_2, \ldots, \ell_r \in \mathbb{F}_p$ and $r, k_1, k_2, \ldots, k_r \in \mathbb{Z}^+$. Each users private n_a and n_b are then elements of the above form (5). As to the subgroups $A, B \leq M \rtimes S$, which must **E**–commute for the protocol to succeed, one possibility is to proceed as follows. Fix an element $z \in M \rtimes S$ and choose two subsets of generators of $M \rtimes S$,

$$\left\{(x_{a_1}(t), s_{a_1}), \ldots, (x_{a_\mu}(t), s_{a_\mu})\right\}, \qquad \left\{(x_{b_1}(t), s_{b_1}), \ldots, (x_{b_\nu}(t), s_{b_\nu})\right\},$$

so that

(6) $\qquad x_{a_i}(t) \cdot x_{b_j}(t) = x_{b_j}(t) \cdot x_{a_i}(t), \qquad i = 1, \ldots, \mu, \quad j = 1, \ldots, \nu,$

(7) $\qquad s_{a_i} t_{b_j} = t_{b_j} s_{a_i}, \qquad i = 1, \ldots, \mu, \quad j = 1, \ldots, \nu,$

and

(8) $\qquad {}^{s_{a_i}} x_{b_j}(t) = x_{b_j}(t), \quad {}^{s_{b_j}} x_{a_i}(t) = x_{a_i}(t), \quad i = 1, \ldots, \mu, \quad j = 1, \ldots, \nu.$

Alice and Bob are then assigned the subgroups

(9) $$\begin{aligned} z \cdot \left\langle (x_{a_1}(t), s_{a_1}), \ldots, (x_{a_\mu}(t), s_{a_\mu}) \right\rangle \cdot z^{-1}, \\ z \cdot \left\langle (x_{b_1}(t), s_{b_1}), \ldots, (x_{b_\nu}(t), s_{b_\nu}) \right\rangle \cdot z^{-1}, \end{aligned}$$

respectively, which will automatically **E**–commute with each other.

Hidden Elements Assumption: *We assume that the element $z \in M \rtimes S$ and the elements $x_{a_1}(t), \ldots, x_{a_\mu}(t), x_{b_1}(t), \ldots, x_{b_\nu}(t) \in M$ are secretly chosen and that it is difficult to determine these elements given that the conjugates (9) are publically announced.*

We are now in a position to summarize the above example of the Algebraic Eraser^{TM} key agreement protocol.

General Public Information: A subgroup M of the matrix group

$$N = GL\bigl(n, \mathbb{F}_p(t_1, \ldots, t_n)\bigr).$$

The symmetric group $S = S_n$ acting on the n variables $t_1, \ldots, t_n$ by permuting them. The subgroup M is chosen to be invariant under the action of S allowing for the formation of the semidirect product $M \rtimes S$.

Covert Information: A finite set of generators,

$$\left\{(x_{a_1}(t), s_{a_1}), \ldots, (x_{a_\mu}(t), s_{a_\mu})\right\} \bigcup \left\{(x_{b_1}(t), s_{b_1}), \ldots, (x_{b_\nu}(t), s_{b_\nu})\right\} \\ \subseteq \left\{(x_1(t), s_1), \ldots, (x_\lambda(t), s_\lambda)\right\},$$

of $M \rtimes S$ satisfying (6), (7), (8), and the hidden elements assumption. An element $z \in M \rtimes S$ satisfying the hidden elements assumption.

Public Information: *An integer $n \geq 7$. A prime number $p > n$. The **E**–commuting subgroups*

$$A = z \cdot \left\langle (x_{a_1}(t), s_{a_1}), \ldots, (x_{a_\mu}(t), s_{a_\mu}) \right\rangle \cdot z^{-1},$$

$$B = z \cdot \left\langle (x_{b_1}(t), s_{b_1}), \ldots, (x_{b_\nu}(t), s_{b_\nu}) \right\rangle \cdot z^{-1},$$

where z is the hidden conjugating element and the $x_i(t)$ are the hidden subgroup generators. The homomorphism $\Pi : M \to N$ satisfying Assumption τ. The operation $\star$ satisfying (2). A fixed matrix $m_0 \in N$.

Alice's Private Key: *A matrix of the form* $n_a = \ell_1 m_0^{\alpha_1} + \ell_2 m_0^{\alpha_2} + \cdots + \ell_r m_0^{\alpha_r}$, *where* $\ell_1, \ldots, \ell_r \in \mathbb{F}_p$ *and* $r, \alpha_1, \ldots, \alpha_r \in \mathbb{Z}^+$ *are secret. A subset of generators* $\left\{ \left(x_{a_{i_1}}(t), s_{a_{i_1}}\right), \ldots, \left(x_{a_{i_\mu}}(t), s_{a_{i_\mu}}\right) \right\}$ *of* A.

Alice's Public Key:
$$A_{\text{Public}} = \left(\left(\cdots \left((n_a, \text{id}) \star z \right) \star \left(x_{a_{i_1}}(t), s_{a_{i_1}}\right) \star \cdots \right) \star \left(x_{a_{i_\mu}}(t), s_{a_{i_\mu}}\right) \right) \star z^{-1}$$

Bob's Private Key: *A matrix of the form* $n_b = \ell'_1 m_0^{\beta_1} + \ell'_2 m_0^{\beta_2} + \cdots + \ell'_{r'} m_0^{\beta_{r'}}$, *where* $\ell'_1, \ldots, \ell'_{r'} \in \mathbb{F}_p$ *and* $r', \beta_1, \ldots, \beta_{r'} \in \mathbb{Z}^+$ *are secret. A subset of generators* $\left\{ \left(x_{b_{j_1}}(t), s_{b_{j_1}}\right), \ldots, \left(x_{b_{j_\nu}}(t), s_{b_{j_\nu}}\right) \right\}$ *of* B.

Bob's Public Key:
$$B_{\text{Public}} = \left(\left(\cdots \left((n_b, \text{id}) \star z \right) \star \left(x_{b_{j_1}}(t), s_{b_{j_1}}\right) \star \cdots \right) \star \left(x_{b_{j_\nu}}(t), s_{b_{j_\nu}}\right) \right) \star z^{-1}$$

Shared Secret:

$$\left(\cdots \left((n_a, \text{id}) \cdot B_{\text{Public}} \star z \right) \star \left(x_{a_{i_1}}(t), s_{a_{i_1}}\right) \right) \star \cdots \right) \star \left(x_{a_{i_\mu}}(t), s_{a_{i_\mu}}\right) \right) \star z^{-1}$$
$$= \left(\cdots \left((n_b, \text{id}) \cdot A_{\text{Public}} \star z \right) \star \left(x_{b_{j_1}}(t), s_{b_{j_1}}\right) \right) \star \cdots \right) \star \left(x_{b_{j_\nu}}(t), s_{b_{j_\nu}}\right) \right) \star z^{-1}.$$

In order to analyze the cryptographic applicability of the above algorithm, we shall make the following simplifying assumptions and definition.

- $i_\mu = j_\nu = g =$ the number of generators in Alice and Bob's private keys.
- $\lambda \leq n^2$ where λ is equal to the number of generators of $M \rtimes S$.

It is now possible to compute the size (in bits) of the public and private keys that occur in the Algebraic Eraser^{TM} Key Agreement Protocol. First of all, Alice and Bob's public keys will simply be a pair consisting of an $n \times n$ matrix with coefficients in the finite field $\mathbb{F}_p$ and an element of the permutation group S_n. Each entry in this matrix will have at most $\log_2(p)$ bits. It follows that the matrix component of the public key will have bit size equal to $n^2 \log_2(p)$. The permuation can be specified by a list of n numbers where each number is between 1 and n. Thus the bit size of the permutation is at most $n \log_2(n) \leq n \log_2(p)$. Consequently, the size of the public key is at most $(n^2 + n) \log_2(p)$. The private key also has two saperate components. First, the high power of the fixed matrix m_0 can be represented with at most $n^2 \log_2(p)$ bits. Secondly, each generator can be specified with at most $\log_2(\lambda) \leq 2\log_2(n)$ bits. It follows that the size of the private key is at most $n^2 \log_2(p) + 2\log_2(n)g$. We record these observations in the following proposition.

Proposition 2. *In the Algebraic Eraser^{TM} protocol specified above, the bit-size of the private key is at most*

$$n^2 \log_2(p) + 2\log_2(n)g, \tag{10}$$

while the bit-size of the public key is at most

$$(n^2 + n)\log_2(p). \tag{11}$$

Next, we examine the running time of the algorithm and show that it is essentially linear in the number of generators g in the private key. We shall obtain a crude estimate of the running time in terms of elementary processor operation. By an elementary processor operation we mean either a search and replace operation or a multiplication/addition/subtraction/involving two bits. It is convenient to make the simplifying assumption that each matrix $x_k(t)$ ($k = 1, 2, \ldots, \lambda$) occurring in the generator $(x_k(t), s_k)$ differs from the identity matrix in at most ℓ entries and that each of these entries is a Laurent polynomial in $\mathbb{F}_p(t)$ where the Laurent polynomial itself has at most ρ terms of degree at most d. For example, the matrix

$$\begin{pmatrix} 1 & 0 & 0 & 0 \\ 0 & t_1 - 2t_2^{-1} - t_2 + t_2^3 & 0 & 0 \\ 0 & 0 & 1 & 0 \\ 3 + 2t_1 & 0 & 0 & 2 \end{pmatrix}$$

differs from the identity matrix in exactly 3 entries and each of these entries involves Laurent polynomials of at most 4 terms of degree at most 3. The degree is defined to be the absolute value of the largest power, i.e., t_2^{-4} has degree 4. Given an element

$$(x(t), s) \in M \rtimes S$$

and a generator

$$(x_j(t), s_j)$$

of $M \rtimes S$, the most expensive and time consuming operation of the protocol is the computation of

$$(x(t), s) \star (x_j(t), s_j) = \left(x(t) \cdot \Pi\left(^s x_j(t)\right),\, ss_j \right).$$

First of all, the multiplication of permutations ss_j can be done in n search and replace operations, so this is clearly linear in the number of generators g. Second, the computation of $^s x_j(t)$ requires at most $\ell\rho$ search and replace operations. The computation of $\Pi\left(^s x_j(t)\right)$ requires an additional $\ell\rho$ search and replace operations followed by at most $\ell\rho d$ computations in $\mathbb{F}_p$. Finally, the computation of $x(t) \cdot \Pi\left(^s x_j(t)\right)$ involves at most $n\ell$ multiplications and additions in $\mathbb{F}_p$. This gives an upper bound of $n + 2\ell\rho + 2\ell\rho d(\log_2 p)^2 + 2n\ell(\log_2 p)^2$ elementary operations for each of the g generators. In the final step of the key agreement protocol, it is necessary to multiply two $n \times n$ matrices over $\mathbb{F}_p$. This will take $n^3(\log_2 p)^2$ operations. Assuming that the conjugating element z is made up of g_z generators, the total estimate for the running time of the algorithm is:

$$n^3(\log_2 p)^2 + (g + 2g_z) \cdot \left(n + 2\ell\rho + 2\ell\rho d(\log_2 p)^2 + 2n\ell(\log_2 p)^2 \right). \tag{12}$$

One may also give estimates for the memory size (fixed and rewriteable) needed to run the protocol.

§5. The Colored Burau Key Agreement Protocol (CBKAP):

Fix an integer $n \geq 7$, and let $t = (t_1, \ldots, t_n)$. Define

$$x_1(t) = \begin{pmatrix} -t_1 & 1 & & & \\ & 1 & & & \\ & & \ddots & & \\ & & & & 1 \end{pmatrix},$$

and for $i = 2, \ldots n-1$, let

$$x_i(t) := \begin{pmatrix} 1 & & & & \\ & \ddots & & & \\ & & t_i & -t_i & 1 \\ & & & \ddots & \\ & & & & 1 \end{pmatrix}.$$

which is the identity matrix except for the ith row where it has successive entries $t_i, -t_i, 1$ with $-t_i$ on the diagonal. For each $i = 1, 2, \ldots, n-1$, we define

$$s_i = (i \ \ i+1)$$

which is just the transposition (element of the symmetric group S_n) which interchanges i and $i+1$. The elements $(x_i(t), s_i)$, for $i = 1, 2, \ldots, n-1$, satisfy the braid relations and hence determine a representation of the braid group (see [AAG2]). Next, fix a prime $p > n$, then the set of pairs

$$\left\{ (x_1(t), s_1), \ldots, (x_{n-1}(t), s_{n-1}) \right\}$$

will generate the semidirect product $M \rtimes S$ with $S = S_n$ and $M \subset GL(n, \mathbb{F}_p)$. We call the group $M \rtimes S$ the colored Burau group. The general key agreement protocol given in §4, with this choice of M, is termed the colored Burau key agreement protocol (CBKAP). If we choose $\tau = (\tau_1, \tau_2, \ldots, \tau_n)$ with $1 \leq \tau_i < p$ for $1 \leq i \leq n$ then one may easily check that Assumption τ of §4 is satisfied.

In order to implement the CBKAP with the above choice of M it is necessary to effectively choose the matrix m_0, the elements $z \in M \rtimes S$, and

$$x_{a_1}(t), \ldots, x_{a_\mu}(t), \ x_{b_1}(t), \ldots, x_{b_\nu}(t) \in M.$$

With regard to the matrix m_0 one can begin by generating a random matrix from $GL(N, \mathbb{F}_p)$ and test to see if this matrix has an irreducible characteristic polynomial over $\mathbb{F}_p$. If it does not we simply choose another random matrix and repeat the process. Appendix A contains a Mathematica program that performs this task which heuristically runs quickly and is always successful. The resulting matrix m_0 then has an easily calculable multiplicative order because m_0 is diagonalizable over $\overline{\mathbb{F}_p}$. The non-zero entries of the diagonal matrix will lie in $\mathbb{F}_{p^n}$ and be the roots of the characteristic polynomial. With pobability better than $1/2$, each of these roots will have order $p^n - 1$, and so the matrix m_0 will likewise have order $p^n - 1$. If the roots have a lower order, we again discard and choose a new m_0. Eventually a suitable m_0 will be found.

We now turn to the task of choosing the elements $z \in M \rtimes S$ and

$$x_{a_1}(t), \ldots, x_{a_\mu}(t),\ x_{b_1}(t), \ldots, x_{b_\nu}(t)\ \in\ M.$$

Assuming that we do not want either party to be able to obtain the other's key, a trusted third party (TTP) will be performing the algorithm. If one wishes to design a system which allows for a "master key" then the TTP would simply be one of the users who would then be in possession of the "master key."

The TTP performs the following actions to establish two commuting sets of generators in the braid group. By the representation described above, this produces two **E**–commuting sets of generators in the colored Burau group. Note that these two sets can then be made public and used by any two parties that wish to establish, secretly, a common key. Thus the TTP need only be called upon once.

Let $B_n = \{b_1, \ldots, b_{n-1}\}$ be the Artin representation of the braid group on n strings. Recall that the left canonical form of a braid word may be written as a power of the fundamental braid times a sequence of short braid words, called permutation braids. For further details see [B]. To further shorten the lengths of keys, any even power of the fundamental braid can be ommitted since it is a central element. For the same reason, any odd power of the fundamental braid can simply be replaced by the fundamental braid itself. This will considerably shorten the sequences of integers representing keys.

TTP Algorithm:

(1) *Choose two secret subsets $BL = \{b_{\ell_1}, \ldots b_{\ell_\alpha}\}$, $BR = \{b_{r_1}, \ldots b_{r_\beta}\}$ of the set of*

generators of B_n, where $|\ell_i - r_j| \geq 2$ for all $1 \leq i \leq \ell_\alpha$ and $1 \leq j \leq r_\beta$.

(2) *Choose a secret element $z \in B_n$.*

(3) *Choose words $\{w_1, \ldots, w_\gamma\}$ of bounded length from BL.*

(4) *Choose words $\{v_1, \ldots, v_\gamma\}$ of bounded length from BR.*

(5) *For $1 \leq i \leq \gamma$:*

  (a) *calculate the left normal form of zw_iz^{-1} and reduce the result modulo the square of the fundamental braid;*

  (b) *set w'_i equal to the sequence of integers that corresponds to the element calculated in (a);*

  (c) *calculate the left normal form of zv_iz^{-1} and reduce the result modulo the square of the fundamental braid;*

  (d) *set v'_i equal to the sequence of integers that corresponds to the element calculated in (c).*

(5) *Publish the two sets $\{w'_1, \ldots, w'_\gamma\}$ and $\{v'_1, \ldots, v'_\gamma\}$.*

§6. Linear Algebraic Attack on CBKAP:

There is a successful linear algebraic attack on CBKAP if the conjugating element z is known. We assume, for simplicity, that n is even. There is a similar attack if n is odd. Suppose that the matrix m_0, the element z, and the user public keys are given by

$$(m_0^\alpha \cdot \Pi(z) \cdot \Pi(^{s_z}A) \cdot \Pi(^{s_z s_A}z^{-1}),\ s_z s_A s_{z^{-1}}),$$

and

$$(m_0^\beta \cdot \Pi(z) \cdot \Pi(^{s_z}B) \cdot \Pi(^{s_z s_B}z^{-1}),\ s_z s_B s_{z^{-1}}).$$

Apriori, the matrix $\Pi(^{s_z}A)$ takes the form $\begin{pmatrix} X & 0 \\ 0 & I \end{pmatrix}$, where X is an element of $GL(n/2, \mathbb{F}_p)$ and I is the identity matrix in $GL(n/2, \mathbb{F}_p)$. Similarly $\Pi(^{s_z}B)$ takes the form $\begin{pmatrix} I & 0 \\ 0 & Y \end{pmatrix}$. Note that, in general, the condition that A, B E-commute require that $\Pi(^{s_z}A)$ and $\Pi(^{s_z}B)$ should be commuting matrices which differ from the identity in disjoint blocks. We can always bring them to the form specified above by conjugating by a suitable permutation matrix.

The attack that emerges does not derive the users' secret keys, but only the agreed upon key. The attacker, Eve, begins by diagonalizing the matrix m_0,

$$Q m_0 Q^{-1} = \begin{pmatrix} \lambda_1 & & \\ & \ddots & \\ & & \lambda_n \end{pmatrix}.$$

Despite the fact that Eve does not know α, she can assert that

$$m_0^\alpha = Q^{-1} \begin{pmatrix} \lambda_1^\alpha & & \\ & \ddots & \\ & & \lambda_n^\alpha \end{pmatrix} Q,$$

and hence Alice's public key actually takes the form

$$(13) \qquad Q^{-1} \begin{pmatrix} \lambda_1^\alpha & & \\ & \ddots & \\ & & \lambda_n^\alpha \end{pmatrix} Q \cdot \Pi(z) \begin{pmatrix} X & 0 \\ 0 & I \end{pmatrix} \Pi(^{s_z s_A}z^{-1}).$$

Since the element z is known Eve can compute $\Pi(z)$, s_z, $s_z s_A$, and then $\Pi(^{s_z s_A}z^{-1})$. Multiplying (13) appropriately on the left and right Eve obtains

$$\begin{pmatrix} \lambda_1^\alpha & & \\ & \ddots & \\ & & \lambda_n^\alpha \end{pmatrix} Q \cdot \Pi(z) \begin{pmatrix} X & 0 \\ 0 & I \end{pmatrix}.$$

Eve may set $Q \cdot \Pi(z) = ((c_{ij}))$, where c_{ij} are known for all $1 \leq i \leq n$, $1 \leq j \leq n$. What remains is simply

$$\begin{pmatrix} \lambda_1^\alpha & & \\ & \ddots & \\ & & \lambda_n^\alpha \end{pmatrix} \cdot ((c_{ij})) \cdot \begin{pmatrix} X & 0 \\ 0 & I \end{pmatrix}.$$

The powers λ_j^α (with $j > \frac{n}{2}$) of the eigenvalues are thus visible. Similarly, one may switch the first half of the columns of Q with the second half of the columns (by conjugating everything by an appropriate permutation matrix) which then allows Eve to obtain m_0^α.

At this point Eve can recover $\Pi(^{s_z}A)$ and the matrix $\Pi(^{s_z s_B s_A} z^{-1})$. The shared key can be obtained from Bob's public key by multiplying on the left and right by known elements: m_0^α on the left, and $(\Pi(^{s_z s_B} z^{-1}))^{-1}$, $\Pi(^{s_z} A)$, and $\Pi(^{s_z s_B s_A} z^{-1})$ on the right (recall that we assumed $\Pi(^{s_z s_B} A) = \Pi(^{s_z} A)$).

The attack just described was implemented in C++. Assuming the conjugating element z in CBKAP is known, the shared secret can be found from the public keys in 0.04 seconds, regardless of the length of the second private keys chosen by Alice and Bob. Part of the C++ code is included in Appendix B. Special thanks to Alan Silvester for producing this code.

§7. Security Analysis of CBKAP:

The linear algebraic attack discussed in §6 cannot be implemented unless the conjugating element z is correctly guessed. This implies that performing an exhaustive search for z can yield Alice's and Bob's shared secret. Let g_z denote the number of Artin generators of z. To ensure m bits of security against this attack we need

(14) $$g_z \cdot \frac{\ln(2n-2)}{\ln(2)} \geq m.$$

Since the advent of braid group cryptography in [AAG1], various algorithms have emerged for determining an element z in the Artin braid group B_n provided the set of elements

$$a_1, \ldots, a_\ell, z^{-1} a_1 z, \ldots, z^{-1} a_\ell z \in B_n$$

is publicly known. This is called the simultaneous conjugacy search problem. We shall examine several such algorithms and show that they cannot successfully generate an attack on the CBKAP.

In [G], a probabilistic approach to the conjugacy search problem in Garside groups is obtained. His algorithm yields a linear solution to the conjugacy search problem in braid groups which almost always works (except on a set of measure zero).

Recall that in the TTP algorithm, the elements

$$w_i' = z w_i z^{-1}, \quad v_j' = z v_j z^{-1}, \qquad (1 \leq i \leq \ell_\alpha, \ 1 \leq j \leq \ell_\beta)$$

are publicly announced, but $z, w_i, v_j, (1 \leq i \leq \ell_\alpha, 1 \leq j \leq \ell_\beta)$ are kept secret. All that we really know is that all the w_i commute with all the v_j, for $1 \leq i \leq \ell_\alpha, 1 \leq j \leq \ell_\beta$. The difficulty in using this attack on CBKAP to recover z, is that although each w_i and v_i is conjugate to w_i' and v_i' respectively, modulo the square of the fundamental braid, w_i and v_i are secrets known only to the TTP. Thus an attacker would have to guess each w_i and v_i correctly and then determine the braid element that conjugates the w_i's and v_i's to the w_i'''s and v_i'''s. The length of each w_i and v_i could easily be set long enough to prevent such an attack but even this is unnecessary for several reasons.

First there will be many elements conjugate to w_1', even of the same length as w_1 if this length was correctly guessed by the attacker. For example all cyclic conjugates of w_1 will generally have the same length as w_1. Second the length of w_1 is only known to be less than a given bound which increases the number of elements conjugate to w_1' which are not w_1, i.e the number of false positives. Third, w_i' was reduced modulo the square of the fundamental braid, further increasing the number of false positives. Finally, each false positive will yield a corresponding false positive for each of the other w_i'''s and v_i'''s.

The reason for the false positives is that the TTP conjugates each by z and then removes the highest power of Δ^2 that he can. In general, after conjugating, you will get much higher powers of Δ^2 in the normal form. Even more importantly, the permutation braids resuting will be all different. For example, if the attacker: chooses words of length 10, then removes powers of Δ^2, and then tries to solve the conjugacy search problem, it will fail to give a usable z.

Length based attacks for the conjugacy search problem were introduced in [HT]. The algorithm has been further studied and developed in [GKTTV], and [D].

The length attacks work as follows. Let $x, y \in B_n$ such that x and y are known to be conjugate in B_n and y is much longer than x. For each $c \in B_n$ of length 10, if the length of $c^{-1}yc$ is significantly less than that of y then to some probability we know that c is the first 10 generators of an element that conjugates x to y. The attacker then repeats the algorithm with $c^{-1}yc$ in place of x.

This attack is ineffective against CBKAP again because the w_i's and v_i's are not known. Further, since w_i' is actually equal to $zw_iz^{-1} \times \Delta^{2r}$, where r is some integer and Δ is the fundamental braid, the actual length of the conjugate element is unknown to all but the TTP. Also, the short length of z as will be determined in the next section, implies that checking all c of length 10, amounts essentially to brute force search for z with only a probabilistic verification of success. Finally if the algorithm is attempted using shorter c's then heuristics indicate that even choosing the correct c produces $c^{-1}yc$ of length equal to or even greater than that of y.

§8. Parameter Choices and Running Times:

The intended application of CBKAP is for constrained devices where only lightweight security is possible. With this, in mind, we will choose parameters to ensure 60, 80, and 120 bits of security, i.e., that, respectively, 2^{60}, 2^{80}, or 2^{120} attempts are needed by any of the above attacks in order to compromise CBKAP. It should be noted however, that CBKAP could be scaled to achieve higher levels of security. We will show the derivation of parameters only for 80 bits of security, the other cases being analogous. The results for all three security levels with a small selection of parameter choices are reported in the tables below. Although it is not the focus of this paper, we note that CBKAP will run at least 10 times faster (but probably significantly more) when implemented in a symmetric platform.

The most immediate and straightforward attack on CBKAP is to perform an exhaustive search of all possible choices of Alice and Bob's private keys. This attack, as one might expect, is easy to overcome by choosing large enough parameters. Recall that Alice's public key is the eraser product of a linear combination, over $\mathbb{F}_p$, of r powers of the publicly known matrix, m_0 and the elements of her second private key which we assume has length g. We know m_0 has order $p^n - 1$ so on average $\frac{1}{2}(p^n - 1)^r$ powers will need to be checked before finding the correct one. Likewise the keyspace for Alice's second private key is all words of length g in generators and their inverses that the TTP gives her. Say there are T of these. Then the total number of searches for her public key will be $\frac{1}{2}(p^n - 1)^r$ for the first private key and T^g for the second private key. Thus for m bits of security, we need $\frac{r \cdot n \cdot \ln(p)}{\ln(2)} \geq m$, and

$$(15) \qquad g \cdot \frac{\ln(T)}{\ln(2)} \geq m.$$

Note that the inequality (15) is critical for without it in place the system is not secure. We will show in the next section that we can choose relatively small parameters to meet our desired standards of security.

The case of $n = 14$ serves as a good illustrative example. Thus CBKAP will run, in this instance, with elements from B_{14} and over matrices of dimension 14. Recall that g_z denotes the length of the element z in the Artin generators. To provide 80 bits of security against an exhaustive search for z, the inequality, as in (14), asserts that we must have

$$g_z \frac{\ln(2n - 2)}{\ln(2)} \geq m$$

implies that $g_z \geq 80 \frac{\ln(2)}{\ln(26)} \approx 17$, and, hence, z must be chosen to be of length 17. If we choose the length of the w_i's and v_i's to be bounded by 10, then the conjugate elements, reduced modulo Δ^2 appear to have length around 188 generators. We label the average length of the w_i's and v_j's by L. Once the length g of Alice's and Bob's second private keys are set, and the prime p is chosen, the number of bit operations required for either of them to generate the shared secret using their private keys and the other's public key is approximately

$$\left(7 \cdot 14 \cdot L \cdot g + 14^2\right) \log_2(p).$$

If the TTP produces 27 generators each for Alice and Bob then including their inverses we have 54 generators in all. Against exhaustive search for either Alice's or Bob's second private keys, we need to ensure, as in (15), that

$$g \cdot \frac{\ln(54)}{\ln(2)} \geq 80.$$

The smallest value possible, satisfying this inequality, is $g = 14$. Likewise, to ensure 80 bits of security against a brute force attack on the first private key, we can choose $p = 13$, and $r = 3$. The chart below summarizes these results and lists the approximate number bit operations needed to produce a shared secret, for the different security levels and the different average lengths of the w_i's and v_j's.

Security Analysis for CBKAP over B_{14}

Bits of Security	Lengths of w_i, v_j	Length of 2nd Private Key	Bit Ops to make shared secret
60	188	11	816144
80	188	14	1037232
120	572	21	4714192

To decrease the number of bit ops, the system can be scaled down further. We next consider $n = 12$. Thus CBKAP will run, in this instance, with elements from B_{12} and over matrices of dimension 14. To provide 80 bits of security against an exhaustive search for z, the inequality

$$g_z \frac{\ln(2n-2)}{\ln(2)} \geq m$$

implies that $g_z \geq 80 \frac{\ln(2)}{\ln(22)} \approx 18$ so z will need to be of length 18. If we choose the length of the w_i's and v_j's to be bounded by 10, then the conjugate elements, reduced modulo Δ^2 appear to have length around 130 generators. We label the average length of the w_i's and v_i's by L. Once the length g of Alice's and Bob's second private keys are set, and the prime p is chosen, the number of bit operations required for either of them to generate the shared secret using their private keys and the other's public key is approximately

$$\left(7 \cdot 12 \cdot L \cdot g + 12^2\right) \log_2 p.$$

If the TTP produces 27 generators each for Alice and Bob then including their inverses we have 54 generators. Against exhaustive search for either Alice's or Bob's second private keys, we need to ensure that

$$g \cdot \frac{\ln(54)}{\ln(2)} \geq 80.$$

The smallest value possible, satisfying this inequality, is $g = 14$. Likewise, to ensure 80 bits of security against a brute force attack on the first private key, we can choose $p = 13$, and $r = 3$. The chart below summarizes these results and lists the approximate number bit operations needed to produce a shared secret, for the different security levels and the different average lengths of the w_i's and v_j's.

Security Analysis for CBKAP over B_{12}

Bits of Security	Lengths of w_i, v_j	Length of 2nd Private Key	Bit Ops to make shared secret
60	130	11	484512
80	188	14	615552
120	572	21	9121312

§9. Hardware Implementations:

The CBKAP may be employed in the RFID space. An initial application would be authentication for a local area network consisting of passive RFID tags and readers. Passive RFID tags are tags with no battery power that draw energy from a reader. We are interested in providing authentication between tag and reader. The CBKAP algorithm allows the RFID tag and a reader to establish a common secret key which can then be used for mutual authentication by standard cryptographic techniques which we do not focus on. We also note that the readers in the local area network typically have heavy weight computational resources and do not require lightweight cryptographic algorithms. Also, the TTP algorithm can be performed offline or by the readers.

We now discuss the part of the CBKAP algorithm which needs to be performed by the passive RFID tag. Consider CBKAP over B_{12} with a security level of 2^{60} and parameters as specified in section 8. For references regarding VLSI signal processing (see [DR]) and for references regarding EPC TM Radio-Frequency Identity Protocols Class (see [EPC]).

Acknowledgment: We would like to thank Mike McGregor for contributing the following last section of this paper.

The practical deployment of the Algebraic Eraser (AE) within an extremely constrained device such as a passive RFID tag is primarily governed by the twin concerns of tag economics and tag performance. At the time of writing these economic constraints limit the total cost of implementation of the AE to less than 0.5 cents. Similarly the performance constraints impose a run time limit of under 20ms if the AE is not to interfere with normal tag access rates provided for by the Electronic Product Code (EPC) protocol or conflict with FCC dwell time regulations. Current CMOS semiconductor industry norms of $1000 per wafer thus limit the available die size for the AE to 163,000 um^2.

One approach is to build a custom Algebraic Eraser processing engine (AEPE) directly in hardware such that the datapaths and dedicated finite field processing elements (Multiplies/Adds) are arranged to compute just enough information on every clock cycle. The AEPE consists of the following four components: A non-volatile memory (NVM) to contain the tag private keys, A permutation engine (SE), A matrix multiplication engine (ME) and control functions (C) to schedule the AE operations. We seek an implementation that balances both sets of constraints

We are now in a position to analyze the following example where $p = 13$ and the AEPE is required to perform 2,200 rounds of a permutation followed by a 12×12 matrix multiplication running at a typical clock rate of 1 MHz.

The matrix multiplier takes advantage of the sparse form of the right hand matrix to reduce the number of computations per row to two finite field multiplications and additions. Furthermore, the choice of $p = 13$ limits all of the operands to 4 bits or less in width. Thus a complete matrix multiplication only requires a total of 24 multiplications and additions. By allowing 5 clock cycles for the matrix multiplication (to give a total run time of 11 ms) we can construct a systolic array multiplier architecture using bit-serial signal processing to reduce all of the datapaths from 4 bits down to 1 bit wide and use the same gates to perform both the multiplication and the addition (the control scheduler allocates 4 clock cycles to the multiplication and 1 to the addition).

The permutation engine SE can be implemented using a Benes switch which, for a bit-serial non-blocking $12 \to 12$ permutation requires less than 56 1-bit wide crossovers or 112 gates.

Having established that we can use 1 bit wide datapaths we complete our gate count and area analysis to determine that we need 5,374 logic gates and 1200 bit of NVM. Based on industry norms for 0.13um CMOS of 242,000 gates/mm2 these functions can be built in circa 22,000 um^2 for the logic and 120,000 um^2 for the NVM.

The gate counts associated with this bit serial architecture are:

Multiplication/Addition = $(24 \times 4 \times 18) = 1728$ gates
Addition = 0 gates (now part of the multiplier).
Inline storage for temporary results = $144 \times (4 \times 4 + 1) = 2448$ gates

AEPE Scheduler < 1200 gates.

Finally we mention that the creation of security standards for the RFID market is at an early stage of development [PKK]

References

[AAG1] Anshel, Iris, Anshel, Michael, Goldfeld, Dorian, *An algebraic method for public-key cryptography*, Math. Res. Lett. 6 (1999), no. 3-4, 287–291.

[AAG2] Anshel, Iris, Anshel, Michael, Goldfeld, Dorian, *A Linear time matrix key agreement protocol over small finite fields*, to appear.

[B] Birman, Joan S. *Braids, links, and mapping class groups*, Annals of Mathematics Studies, No. 82. Princeton University Press, Princeton, N.J.; University of Tokyo Press, Tokyo, 1974.

[D] Dehornoy, Patrick, *Braid-based cryptography*. Group theory, statistics, and cryptography, 5–33, Contemp. Math., 360, Amer. Math. Soc., Providence, RI, 2004.

[DR] *VLSI Signal Processing - A Bit Serial Approach,* Denyer & Renshaw, (1985), Addison-Wesley, 0-201-14404-2

[EPC] *EPC^{TM} Radio-Frequency Identity Protocols Class-1 Generation-2 UHF RFID Protocol for Communications at 860 - 960 MHz,* Version 1.0.9, EPCGlobal Inc, (2004).

[GKTTV] D. Garber, S. Kaplan, M. Teicher, B. Tsaban, U. Vishne, *Length-based conjugacy search in the Braid group* , math. GR0209267)

[GJP] S.L. Garfinkel, A. Juels and R. Pappu, *RFID Privacy: An Overview of Solutions and Problems,* IEEE Security and Privacy Volume 3 Number 3 (May/June 2005), 34–43.

[G] V. Gebhardt, *A new approach to the conjugacy problem in Garside groups*, (English. English summary) J. Algebra 292 (2005), no. 1, 282–302.

[HT] J. Hughes, A. Tannenbaum, *Length-based attacks for certain group based encryption rewriting systems*, 2000 (Published in SECI 02)).

[KM] N. Koblitz and A.J. Menenzes, *A Survey of Public-Key Cryptosystems*, SIAM Review, Volume 44, Number 4 (December 2005), 599–634.

[MKS] Magnus, Wilhelm, Karrass, Abraham, Solitar, Donald *Combinatorial group theory, Presentations of groups in terms of generators and relations*, Reprint of the 1976 second edition. Dover Publications, Inc., Mineola, NY, 2004.

[OSK] M. Ohkubo, K. Suzuki, and S. Kinoshita,*RFID Privacy Issues and Technical Challenges,* Communications of the ACM. Volume 48, Number 9 (2005), 66–71.

[PKK] T. Phillips, T. Karygiannis, T. Kuhn, *Security Standards for the RFID Market*, IEEE Security & Privacy Volume 3 Number 6 (November/December 2005), 85-89.

IRIS ANSHEL, 31 PETER LYNAS CT. TENAFLY, NJ 07670, USA

MICHAEL ANSHEL, CITY COLLEGE OF NEW YORK, NEW YORK, NY 10031, USA

DORIAN GOLDFELD, COLUMBIA UNIVERSITY, DEPARTMENT OF MATHEMATICS, NEW YORK, NY 10027

STEPHANE LEMIEUX, UNIVERSITY OF CALGARY, DEPARTMENT OF MATHEMATICS, CALGARY, CANADA

Appendix A

Generate a matrix of high order in Maple
By: Stephane Lemieux
Date: August 2005

```
> interface(rtablesize):=13;
```
$$interface\,(rtablesize) := 13$$

This Maple program generates a 12 by 12 matrix with irreducible characteristic polynomial over F_p. Such matrices usually have order $p^{12} - 1$.

```
> matgen:=proc(p::posint)
>     local tag, m, i, j, r, f;
>     tag := 0;
>     while tag = 0 do
>        m:=Matrix(12,12):
>        r:=rand(p):
>        for i from 1 to 12 do
>           for j from 1 to 12 do
>              m[i,j]:=r();
>           od;
>        od;
>        f := x -> Det(m - x*Matrix(12,12,shape=identity))
>              mod p;
>        if Irreduc(f) mod p then return m; fi;
>     od:
> end proc:

> matgen(13);
```

$$\begin{bmatrix}
2 & 5 & 6 & 8 & 6 & 11 & 4 & 7 & 6 & 11 & 0 & 11 \\
8 & 11 & 2 & 7 & 6 & 7 & 8 & 0 & 4 & 5 & 4 & 4 \\
3 & 7 & 0 & 12 & 5 & 5 & 10 & 2 & 9 & 5 & 5 & 1 \\
2 & 5 & 12 & 5 & 0 & 3 & 6 & 0 & 9 & 3 & 9 & 10 \\
2 & 3 & 11 & 7 & 9 & 8 & 6 & 3 & 1 & 12 & 3 & 5 \\
6 & 9 & 0 & 10 & 5 & 6 & 11 & 3 & 0 & 9 & 7 & 5 \\
0 & 9 & 11 & 12 & 8 & 12 & 7 & 1 & 3 & 2 & 4 & 3 \\
12 & 8 & 0 & 3 & 5 & 12 & 0 & 11 & 0 & 2 & 2 & 0 \\
12 & 0 & 0 & 5 & 6 & 8 & 10 & 5 & 7 & 5 & 2 & 7 \\
4 & 6 & 12 & 9 & 6 & 10 & 5 & 4 & 9 & 6 & 7 & 0 \\
0 & 6 & 11 & 12 & 9 & 0 & 7 & 7 & 11 & 9 & 5 & 2 \\
12 & 4 & 4 & 10 & 9 & 5 & 5 & 7 & 9 & 2 & 6 & 5
\end{bmatrix}$$

```
>  matgen(997);
```

$$\begin{bmatrix}
795 & 20 & 774 & 716 & 581 & 13 & 57 & 611 & 231 & 275 & 448 & 637 \\
83 & 480 & 556 & 14 & 927 & 157 & 378 & 968 & 514 & 763 & 165 & 476 \\
631 & 669 & 457 & 748 & 620 & 23 & 854 & 635 & 906 & 28 & 595 & 869 \\
809 & 867 & 517 & 273 & 270 & 81 & 303 & 649 & 768 & 875 & 256 & 446 \\
309 & 840 & 893 & 824 & 50 & 91 & 515 & 436 & 713 & 656 & 658 & 182 \\
209 & 687 & 123 & 97 & 371 & 43 & 351 & 599 & 508 & 799 & 140 & 686 \\
333 & 74 & 298 & 956 & 728 & 47 & 303 & 24 & 294 & 443 & 477 & 195 \\
224 & 333 & 123 & 638 & 428 & 447 & 129 & 712 & 581 & 368 & 728 & 127 \\
605 & 50 & 549 & 164 & 725 & 515 & 974 & 511 & 530 & 70 & 519 & 500 \\
855 & 786 & 856 & 607 & 159 & 522 & 409 & 541 & 773 & 241 & 670 & 767 \\
766 & 437 & 838 & 695 & 542 & 729 & 149 & 213 & 338 & 497 & 673 & 200 \\
953 & 691 & 447 & 407 & 410 & 266 & 100 & 120 & 566 & 222 & 79 & 316
\end{bmatrix}$$

```
>  m:=matgen(13);
```

$$m := \begin{bmatrix}
6 & 0 & 0 & 1 & 1 & 0 & 2 & 1 & 6 & 3 & 9 & 0 \\
6 & 0 & 4 & 7 & 12 & 2 & 7 & 2 & 2 & 11 & 8 & 6 \\
2 & 2 & 8 & 7 & 8 & 9 & 3 & 9 & 3 & 3 & 9 & 10 \\
1 & 0 & 12 & 8 & 0 & 0 & 8 & 7 & 7 & 1 & 10 & 4 \\
6 & 5 & 12 & 2 & 4 & 7 & 0 & 8 & 0 & 8 & 0 & 12 \\
6 & 6 & 5 & 2 & 3 & 5 & 1 & 10 & 0 & 2 & 3 & 1 \\
4 & 2 & 3 & 3 & 12 & 12 & 1 & 5 & 6 & 0 & 1 & 12 \\
12 & 3 & 5 & 7 & 12 & 3 & 6 & 3 & 5 & 2 & 7 & 10 \\
8 & 5 & 2 & 1 & 11 & 7 & 2 & 6 & 5 & 9 & 5 & 8 \\
5 & 4 & 0 & 6 & 12 & 1 & 9 & 3 & 12 & 7 & 11 & 0 \\
0 & 3 & 4 & 6 & 4 & 6 & 7 & 9 & 9 & 0 & 0 & 10 \\
12 & 11 & 9 & 7 & 10 & 8 & 4 & 9 & 2 & 0 & 2 & 5
\end{bmatrix}$$

```
>  Det(m) mod 13;
```
$$0$$
```
>  x:=matgen(5);
```

$$x := \begin{bmatrix} 3 & 4 & 2 & 0 & 0 & 4 & 0 & 3 & 1 & 0 & 0 & 3 \\ 4 & 0 & 1 & 2 & 2 & 2 & 1 & 3 & 0 & 3 & 1 & 3 \\ 3 & 3 & 0 & 1 & 4 & 0 & 0 & 2 & 2 & 2 & 2 & 3 \\ 4 & 0 & 0 & 3 & 1 & 2 & 0 & 1 & 1 & 2 & 3 & 3 \\ 2 & 4 & 0 & 3 & 1 & 1 & 0 & 1 & 1 & 2 & 2 & 3 \\ 1 & 4 & 3 & 0 & 4 & 4 & 4 & 3 & 4 & 2 & 3 & 0 \\ 3 & 0 & 4 & 3 & 4 & 3 & 4 & 0 & 0 & 3 & 1 & 2 \\ 4 & 1 & 4 & 2 & 4 & 1 & 0 & 2 & 2 & 3 & 0 & 1 \\ 1 & 1 & 4 & 1 & 4 & 0 & 2 & 2 & 0 & 4 & 2 & 1 \\ 0 & 4 & 0 & 0 & 1 & 3 & 2 & 4 & 3 & 3 & 0 & 4 \\ 2 & 2 & 1 & 1 & 4 & 3 & 1 & 4 & 2 & 0 & 2 & 1 \\ 3 & 2 & 0 & 2 & 3 & 2 & 3 & 3 & 4 & 1 & 0 & 0 \end{bmatrix}$$

\> Det(x) mod 5;

1

Appendix B

```
/*
 *  Program: attack2
 *  Created: 2005-09-20 Alan Silvester
 *
 *  Implements the attack
 */

#include <iostream>
#include <sstream>
#include <fstream>
#include <string>

#include <NTL/mat_ZZ_p.h>

#include "assert.h"
#include "timer.h"
#include <sys/time.h>

#include "functions.h"

#include "field_functions.h"

#include "AAG_System.h"

NTL_CLIENT

using namespace std;
using namespace NTL;

std::ostream & operator<<(std::ostream &os,
                          const GeneratorIndex &gi)
{
   gi.print(os);
}

Blob AAGSpace::u_compute (const ValueMatrix& seed) const

{
   Blob x (seed, PermutationVector (m_matrix_size, true));
   size_t i;

   for (i = 0; i < m_u.size (); ++i)
```

```
   {
      x = eraser (x, *(get_generator (m_u[i])));
   }

   return (x);
}


Blob AAGSpace::u_inv_compute (const ValueMatrix& seed,
                              const PermutationVector& p1) const
{
   Blob x (seed, p1);

   size_t i;

   for (i = 0; i < m_u_inverse.size (); ++i)

   {
      x = eraser (x, *(get_generator (m_u_inverse[i])));
   }

   return (x);
}

int main (int argc, char * argv[])
{
   //////////////////////////////////////////////////////////
   //
   // Opening output log file
   //
   //////////////////////////////////////////////////////////

   ofstream output ("outputfile3.txt");
   if (!output.is_open())
   {
      cout << "error opening outputfile3.txt\n";
      return 1;
```

```
    }

    // Seed RNG
//  timeval tv;
//  gettimeofday(&tv, NULL);
//  srand(tv.tv_sec);

    /////////////////////////////////////////////////////////
    //
    // Writing temp ini file
    //
    /////////////////////////////////////////////////////////

    int kl = 100;

    ofstream inioutput ("test5.ini");
    if (!inioutput.is_open())
    {
       cout << "error opening test5.ini\n";
       return 1;
    }

//  inioutput << "RandomSeed=" << tv.tv_sec << endl;
    inioutput << "DisplaySharedSecret=no\n";

    inioutput << "Iterations=10\n";

    inioutput << "Prime=13\n";

    inioutput << "VC=12\n";

    inioutput << "T=1D,1 2 8 9 10 11 12 3 4 5 6 7\n";

    inioutput << "KeyLength=" << kl << endl;

    inioutput << "Partition=1D,1 2 3 4 5\n";
    inioutput << "Partition=1D,7 8 9 10 11\n";

    inioutput << "ULength=12\n";
    inioutput << "U=1,12\n";
    inioutput << "U=7,-12\n";
    inioutput << "U=12,-12\n";
    inioutput << "U=8,12\n";
    inioutput << "U=9,-12\n";
    inioutput << "U=11,12\n";
    inioutput << "U=4,-12\n";
```

```
inioutput << "U=5,-12\n";
inioutput << "U=6,12\n";
inioutput << "U=2,-12\n";
inioutput << "U=3,12\n";
inioutput << "U=10,-12\n";
inioutput << "U=1,-12\n";
inioutput << "U=7,12\n";
inioutput << "U=12,12\n";
inioutput << "U=8,-12\n";
inioutput << "U=9,12\n";
inioutput << "U=11,-12\n";
inioutput << "U=4,12\n";
inioutput << "U=5,12\n";
inioutput << "U=6,-12\n";
inioutput << "U=2,12\n";
inioutput << "U=3,-12\n";
inioutput << "U=10,12\n";

inioutput.close();

////////////////////////////////////////////////////////////
//
// Initialisation
//
////////////////////////////////////////////////////////////

// Setup the generator matricies
vector<pair<Generator, Generator> > g;
size_t matrix_size = 6;
timer t;
cout << "setting up generator matricies\n";

cout << "making new AAGSpace\n"; ;
AAGSpace space = AAGSpace("test5.ini");

cout << "have " << space.m_partitions.size()
     << " partitions\n"; ;

cout << "m_partition[0] data: ";
for (int i = 0; i < space.m_partitions[0].size(); i++)
   cout << space.m_partitions[0][i] << " ";
cout << endl; ;

cout << "m_partition[1] data: ";
for (int i = 0; i < space.m_partitions[1].size(); i++)
   cout << space.m_partitions[1][i] << " ";
```

```
cout << endl; ;

ValueMatrix vm_seed_matrix = ValueMatrix(matrix_size*2,
                                         true);

//////////////////////////////////////////////////////////
//
// Start timer for alice key-gen
//
//////////////////////////////////////////////////////////

t.start();

//////////////////////////////////////////////////////////
//
// Generate Alice's keypair
//
//////////////////////////////////////////////////////////

cout << "generating alice's keypair\n";

// This skips the exponentiation of the seed matrix
Keypair alice_key = space.make_keypair(0, 0);

cout << "keygen time: " << t << endl;
Blob alice_key_blob = alice_key.build_public();
alice_key.print("Keypair::print - alice's keypair", cout);

//////////////////////////////////////////////////////////
//
// Start timer for bob key-gen
//
//////////////////////////////////////////////////////////

t.restart();

//////////////////////////////////////////////////////////
//
// Generate Bob's keypair
//
//////////////////////////////////////////////////////////

cout << "generating bob's keypair\n";

// This skips the exponentiation of the seed matrix
Keypair bob_key = space.make_keypair(1, 0);
```

```
cout << "keygen time: " << t << endl;
Blob bob_key_blob = bob_key.build_public();
bob_key.print("Keypair::print - bob's keypair", cout);

////////////////////////////////////////////////////////
//
// Start timer for key exchange
//
////////////////////////////////////////////////////////

t.restart();

////////////////////////////////////////////////////////
//
// Shared secret
//
////////////////////////////////////////////////////////

Blob shared_secret = space.key_exchange(bob_key_blob,
                                        alice_key.PrivateKey(),
                                        vm_seed_matrix);
cout << "secret gen time: " << t << endl;
shared_secret.print("shared secret (alice)", cout);

space.m_t.print("t = ");

cout << "Key exchange time: " << t << endl;

////////////////////////////////////////////////////////
//
// Start timer for crack
//
////////////////////////////////////////////////////////

t.restart();

////////////////////////////////////////////////////////
//
// Make identity matrix
//
////////////////////////////////////////////////////////

ValueMatrix vm = ValueMatrix(matrix_size*2, true);

////////////////////////////////////////////////////////
```

```
//
// Calculate pi(u)
//
////////////////////////////////////////////////////////////

cout << "Computing pi(u)\n";
Blob pi_u = space.u_compute(vm);
pi_u.print("pi(u), u.p", cout);

////////////////////////////////////////////////////////////
//
// Calculate P1 (alice)
//
////////////////////////////////////////////////////////////

// Dummy data
PermutationVector alice_p1 = alice_key_blob.m_permutation;

alice_key_blob.m_permutation.Permute(pi_u.m_permutation,
                                     alice_p1);

cout << "\nP1 (alice)\n";
alice_p1.print(cout);
cout << endl;

////////////////////////////////////////////////////////////
//
// Calculate P1 (bob)
//
////////////////////////////////////////////////////////////

// Dummy data
PermutationVector bob_p1 = bob_key_blob.m_permutation;
bob_key_blob.m_permutation.Permute(pi_u.m_permutation,
                                   bob_p1);

cout << "\nP1 (bob)\n";
bob_p1.print(cout);
cout << endl;

////////////////////////////////////////////////////////////
//
// Calculate pi(u inv) (alice)
//
////////////////////////////////////////////////////////////
```

```
cout << "Computing pi(u inv to sig u sig a)\n";
Blob alice_pi_u_inv = space.u_inv_compute(vm, alice_p1);
alice_pi_u_inv.print("pi(u_inv_to_sig_u_sig_a), alice_p1",
                     cout);

//////////////////////////////////////////////////////////
//
// Calculate pi(u inv) (bob)
//
//////////////////////////////////////////////////////////

cout << "Computing pi(u inv to sig u sig b)\n";
Blob bob_pi_u_inv = space.u_inv_compute(vm, bob_p1);
bob_pi_u_inv.print("pi(u_inv_to_sig_u_sig_b), bob_p1",
                   cout);

//////////////////////////////////////////////////////////
//
// Calculate pi(u inv)
//
//////////////////////////////////////////////////////////

cout << "Computing pi(u inv)\n";
Blob pi_u_inv = space.u_inv_compute(vm,
                   PermutationVector(matrix_size*2, true));
pi_u_inv.print("pi(u inv), sig_u_inv", cout);

//////////////////////////////////////////////////////////
//
// Calculate P2 (alice)
//
//////////////////////////////////////////////////////////

// Dummy data
PermutationVector alice_p2 = alice_key_blob.m_permutation;

pi_u_inv.m_permutation.Permute(alice_p1, alice_p2);

cout << "\nP2 (alice)\n";
alice_p2.print(cout);
cout << endl;

//////////////////////////////////////////////////////////
//
// Calculate P2 (bob)
//
```

```
/////////////////////////////////////////////////////////////

// Dummy data
PermutationVector bob_p2 = bob_key_blob.m_permutation;

pi_u_inv.m_permutation.Permute(bob_p1, bob_p2);

cout << "\nP2 (bob)\n";
bob_p2.print(cout);
cout << endl;

/////////////////////////////////////////////////////////////
//
// Compute pi(u)^(-1)
//
/////////////////////////////////////////////////////////////

// Init ZZ_p class
ZZ p;
p = (long) 13;
ZZ_p::init(p);

// Determinant holder (don't really need)
ZZ_p d1;

// Our matrix and matrix holder
mat_ZZ_p m1;
mat_ZZ_p X;
m1.SetDims(matrix_size*2, matrix_size*2);
X.SetDims(matrix_size*2, matrix_size*2);

// Fill matrix with data
for (int i = 0; i < matrix_size*2; i++)
   for (int j = 0; j < matrix_size*2; j++)
      m1[i][j] = pi_u.m_matrix.cell(i,j);

// Calculate inverse
NTL::inv(d1, X, m1);

// Copy data back out
ValueMatrix inverse_pi_u (matrix_size*2, false);

for (int i = 0; i < matrix_size*2; i++)
   for (int j = 0; j < matrix_size*2; j++)
      for (int k = 0; k < rep(X[i][j]).size(); k++)
         inverse_pi_u.cell(i, j, inverse_pi_u.cell(i,j)
```

```
                                + digit(rep(X[i][j]), k));

cout << "inv of pi(u)\n";
inverse_pi_u.print(cout);

//////////////////////////////////////////////////////////
//
// Compute pi(u^p1)^(-1) (alice)
//
//////////////////////////////////////////////////////////

Blob alice_m2_blob = space.make_m2(vm, alice_p1,
                                   matrix_size);

// Determinant holder (don't really need)
ZZ_p alice_d2;

// Our matrix and matrix holder
mat_ZZ_p alice_m2;
mat_ZZ_p alice_Y;
alice_m2.SetDims(matrix_size*2, matrix_size*2);
alice_Y.SetDims(matrix_size*2, matrix_size*2);

// Fill matrix with data
for (int i = 0; i < matrix_size*2; i++)
   for (int j = 0; j < matrix_size*2; j++)
      alice_m2[i][j] = alice_m2_blob.m_matrix.cell(i,j);

// Calculate inverse
NTL::inv(alice_d2, alice_Y, alice_m2);

// Copy data back out
ValueMatrix alice_m2_inverse (matrix_size*2, false);

for (int i = 0; i < matrix_size*2; i++)
   for (int j = 0; j < matrix_size*2; j++)
      for (int k = 0; k < rep(alice_Y[i][j]).size(); k++)
         alice_m2_inverse.cell(i, j,
                            alice_m2_inverse.cell(i,j)
                            + digit(rep(alice_Y[i][j]), k));

cout << "inv of pi(u_inv_to_sig_u_sig_a)\n";
alice_m2_inverse.print(cout);

//////////////////////////////////////////////////////////
//
```

```
// Compute pi(u^p1)^(-1) (bob)
//
/////////////////////////////////////////////////////////////

Blob bob_m2_blob = space.make_m2(vm, bob_p1, matrix_size);

// Determinant holder (don't really need)
ZZ_p bob_d2;

// Our matrix and matrix holder
mat_ZZ_p bob_m2;
mat_ZZ_p bob_Y;
bob_m2.SetDims(matrix_size*2, matrix_size*2);
bob_Y.SetDims(matrix_size*2, matrix_size*2);

// Fill matrix with data
for (int i = 0; i < matrix_size*2; i++)
   for (int j = 0; j < matrix_size*2; j++)
      bob_m2[i][j] = bob_m2_blob.m_matrix.cell(i,j);

// Calculate inverse
NTL::inv(bob_d2, bob_Y, bob_m2);

// Copy data back out
ValueMatrix bob_m2_inverse (matrix_size*2, false);

for (int i = 0; i < matrix_size*2; i++)
   for (int j = 0; j < matrix_size*2; j++)
      for (int k = 0; k < rep(bob_Y[i][j]).size(); k++)
         bob_m2_inverse.cell(i, j, bob_m2_inverse.cell(i,j)
                             + digit(rep(bob_Y[i][j]), k));

cout << "inv of pi(u_inv_to_sig_u_sig_b)\n";
bob_m2_inverse.print(cout);

/////////////////////////////////////////////////////////////
//
// Recover key
//
/////////////////////////////////////////////////////////////

ValueMatrix alice_temp = (inverse_pi_u
                          * alice_key_blob.m_matrix)
                         * alice_m2_inverse;
cout << "\nShould be upper-block (alice)\n";
alice_temp.print(cout);
```

```
ValueMatrix bob_temp = (inverse_pi_u
                        * bob_key_blob.m_matrix)
                        * bob_m2_inverse;
cout << "\nShould be lower-block (bob)\n";
bob_temp.print(cout);

// Compose sigma_a, sigma_b

// Dummy data
PermutationVector result1 = alice_p1;
PermutationVector result2 = alice_p1;

alice_p1.Permute(bob_p2, result1);
result1.Permute(pi_u_inv.m_permutation, result2);

cout << "\nalice_p1\n";
alice_p1.print(cout);
cout << "\nalice_p2\n";
alice_p2.print(cout);
cout << "\nbob_p1\n";
bob_p1.print(cout);
cout << "\nbob_p2\n";
bob_p2.print(cout);
cout << "\nresult 1\n";
result1.print(cout);
cout << "\nresult 2\n";
result2.print(cout);

Blob rhs = space.u_inv_compute(vm, result1);

ValueMatrix found_secret = pi_u.m_matrix * alice_temp
                           * bob_temp * rhs.m_matrix;
cout << "\n\nsecret is\n";
found_secret.print(cout);
result2.print(cout);

////////////////////////////////////////////////////////
//
// Stop timer and print
//
////////////////////////////////////////////////////////

t.stop();
cout << "\n\nRun-time: " << t << endl;
```

```
////////////////////////////////////////////////////////
//
// Cleanup and exit
//
////////////////////////////////////////////////////////

    return 0;
}
```

Designing Key Transport Protocols Using Combinatorial Group Theory

G.Baumslag, T.Camps, B.Fine, G.Rosenberger, and X.Xu

ABSTRACT. There have been several proposed cryptosystems translating the classical Diffie-Hellman public key exchange protocol into group theory - specifically nonabelian groups. Braid group based cryptography can be considered among these. Here we examine a somewhat simpler approach to developing key transport protocols based on a nonabelian group having either a large abelian subgroup or two large subgroups which commute elementwise. We then give some sample platform groups for use with this system.

1. Introduction

Most common public key cryptosystems and public key exchange protocols presently in use, such as the RSA algorithm, Diffie-Hellman, and elliptic curve methods are number theory based and hence depend on the structure of abelian groups. The strength of computing machinery has made these techniques theoretically susceptible to attack and hence there has been a line of research to develop cryptosystems and key exchange protocols in nonabelian groups. (See [AAG], [D], [BFX 1,2], [KLCHKP], [PHKCP],[SU]). The important sources of nonabelian groups that can be used in cryptosystems are combinatorial group theory and linear group theory. Braid group cryptography, where encryption is done within the classical braid groups, is one prominent example. The one way functions in braid group systems are based on the difficulty of solving group theoretic decision problems such as the conjugacy problem. Although braid group cryptography had initial spectacular success, various potential attacks have been identified. Borovik, Myasnikov, Shpilrain [BMS] and others have studied the statistical aspects of these attacks and have identitifed what are termed black holes in the platform groups outside of which present cryptographic problems. In [BFX 1,2] and [X] potential cryptosystems using a combination of combinatorial group theory and linear groups were suggested and a general schema for the these types of cryptosystems was given. In [BFX 2] a public key version of this schema using the classical modular group as a platform was presented. A cryptosystem using the the extended modular group $SL_2(\mathbb{Z})$ was developed by Yamamura ([Y]) but was subsequently shown to have loopholes

---

2000 *Mathematics Subject Classification.* Primary 94A60; Secondary 20G20, 20H05.
Key words and phrases. Cryptosystem, Free Group, Linear Group, Modular Group, Diffie-Hellman, Braid group cryptography, Key Transport Protocol.

([BG],[S],[HGS]). In [BFX 2] attacks based on these loopholes were closed. It has been suggested that at this point further pure group theoretic research, with an eye towards cryptographic applications, is necessary. In particular, although the present braid group cryptosystems may be attackable the basic group theoretic ideas are important. What is then necessary is then to look at other (nonabelian) group theoretical methods as well as additional potential platform groups. There have been several proposed cryptosystems translating the classical Diffie-Hellman public key exchange protocol into group theory. Braid group based cryptography can be considered among these.

In this paper we examine a somewhat simpler approach based on a nonabelian group having either a large abelian subgroup or two large subgroups which elementwise commute. We then present some sample platform groups for use with this system. In the standard Diffie-Hellman system the roles of the communicating parties, Bob and Alice, are symmetric. In the group theoretic analogue that we propose the secret key is determined by one of the parties, say Bob, and then communicated to the other. Hence our method, although based on Diffie-Hellman, actually belongs to the class of **key transport protocols** (see [BM]).

In the next section we describe this group theoretic key transport method and the necessary conditions that a group must have to be considered as a potential platform group. In section 3 we give five proposed platform groups that match these criteria. The first uses as a secret key an automorphism of a free group. The encryption is done within a free group and the the platform group for the key exchange protocol is $Aut(F_n)$, the automorphism group of the free group. The second proposed platform group is $SL_4(\mathbb{Z})$ and the abelian subgroups are copies of $SL_2(\mathbb{Z})$ given as upper and lower blocks in the whole group. The key is an element of $SL_4(\mathbb{Z})$ and the encryption is done within free subgroups of the whole group. The next proposed class of platform groups are the surface braid groups. Part of this material is in the thesis of Xu [X] and the thesis work of Camps [C]. The final two examples are abstract - the free product of two free groups with commuting subgroups and residually free groups given as iterated HNN extensions of free groups with the free part part centralizing a given large subgroup.

This work can be considered as part of a general plan to determine appropriate algebraic objects (both commutative and noncommutative) that can serve as platforms for cryptosystems. Security assessments of the proposed methods are being considered in the thesis work of Camps [C].

2. A Group Theoretic Key Transport Protocol

Public key exchange protocols can be developed using nonabelian groups. Braid group cryptosystems and key exchange protocols have been developed by Anshel,Anshel,Goldfeld [AAG] and Ko,Lee et.al [KLCHKP]. The article by Dehorny [D] gives a nice discussion of the ideas, both mathematical and cryptographic in these systems. There have been various attacks on the AAG scheme. The key idea in this method is the difficulty of the conjugator search problem in the braid groups. Attacks on this system assume then that the platform group is a braid group. However the key exchange idea suggested by AAG is entirely general and can be applied in a wide variety of platform groups. As suggested by Dehorny and others this calls not for a scrapping of the AAG system but actually calls for further mathematical research on alternate platform groups and evaluations of these platforms relative to the known attacks.

In this paper we use a slightly different method which is a general group theoretic analogue of the classical Diffie-Hellman protocol. Historically this was the first proposed public key exchange. Recall that the Diffie-Hellman method depends on the difficulty of solving the discrete logarithm problem coupled with the fact that exponents commute (see [Ko]). That is given a primitive root $g \mod p$ the problem of finding g given g^e. The technique works as follows. Bob and Alice will use a classical cryptosystem based on a key k with $1 < k < p-1$ where p is a prime. Alice and Bob must communicate this secret key. Recall that the multiplicative group of the finite field Z_p is cyclic and let g be a multiplicative generator. Alice chooses an integer a with $1 < a < p-1$ and makes public g^a. Bob chooses an integer b with $1 < b < p-1$ and makes public g^b. The secret key k is g^{ab}. Both Bob and Alice, but presumably no one else, can discover this key. Alice knows a and g^b is public from Bob. She can compute $g^{ab} = g^{ba} = (g^b)^a$. An analogous situation holds for Bob. An attacker only knows g^a and g^b. Unless the attacker can solve the discrete log problem, that is find the base g, the key exchange is secure.

2.1. The Key Transport Protocol. A group theoretic key transport protocol based on the Diffie-Hellman scheme can be developed in the following manner. Suppose that we have a finitely presented group G that can be represented in a nice way - by this we mean either as a matrix group or as words relative to a nice presentation. Further suppose that G has two large subgroups A_1, A_2 that commute elementwise. Alternatively we could use one large abelian subgroup A of G. The meaning of large is, of course, hazy but here means that within G it is difficult to determine when an arbitrary element is in A (or A_1, A_2) and further A (or A_1, A_2) is large enough so that random choices can be made from them.

Now suppose that Bob wants to communicate with Alice via an open airway. The secret key telling them which encryption system to use is encoded within the finitely generated group G with the properties given above. The two subgroups A_1, A_2 which commute elementwise are kept secret by Bob and Alice. A_1 is the subgroup for Bob and A_2 the subgroup for Alice. Bob wants to send the key $W \in G$ to Alice. He chooses two random elements $B_1, B_2 \in A_1$ and sends Alice the message (in encrypted form) $B_1 W B_2$. Alice now chooses two random elements $C_1, C_2 \in A_2$ and sends $C_1 B_1 W B_2 C_2$ back to Bob. These messages appear in the representation of G and hence for example as matrices or as reduced words in the generators so they don't appear as solely concatenation of letters. Since A_1 commutes elementwise with A_2 we have

$$C_1 B_1 W B_2 C_2 = B_1 C_1 W C_2 B_2.$$

Further since Bob knows his chosen elements B_1 and B_2 he can multiply by their inverses to obtain $C_1 W C_2$ which he then sends back to Alice. Since Alice knows her chosen elements C_1, C_2 she can multiply by their inverses to obtain the key W. It is assumed that for each message Bob and Alice would choose different pairs of random elements from either A_1 or A_2.

Notice that although this is roughly based on the Diffie-Hellman method it is not symmetric in the communicating parties. In the present scheme the secret key is completely determined by Bob, who then communicates it to Alice. The scheme then falls into the class of key transport protocols rather than public key exchange protocols. Key transport protocols are in most cases designed assuming that an underlying encryption system (and usually also a signature verification system) is in place. The security of the key transport protocol will rely on the security of these auxillary schemes (see [BM Chapter 4]).

In the group theoretic proposal, the encryption scheme is suggested to be done within the same group as the key transport protocol alhtough this is not essential. In the group theortic key transport protocol an attacker Eve has knowledge of the overall group G and a view of encrypted messages. The security lies in the difficulty of determining the elementwise commuting subgroups A_1, A_2, which are kept secret by Bob and Alice, and in the security of the actual encryption scheme. Security assessments of the method, relative to the particular proposed platform groups, is being carried out in the dissertation work of Camps [C].

2.2. Encryption Schemes Using Combinatorial Group Theory.

The problem now is in choosing appropriate platform groups and then encryption systems to which this group theoretic key transport protocol can be applied. We do this in the next section. The encryption systems that we propose to use are based on combinatorial group theory so before presenting these platform groups we briefly describe free group cryptography (see also [BFX 1,2] or [X]).

The basic idea in using combinatorial group theory for cryptography is that elements of groups can be expressed as words in some alphabet. If there is an easy method to rewrite group elements in terms of these words and further the technique used in this rewriting process can be supplied by a secret key then a cryptosystem can be created. The books by Baumslag [B], Lyndon and Schupp [LS] and Magnus, Karrass and Solitar [MKS], are standard references for material on combinatorial group theory.

Consider a free group F on free generators $x_1, ..., x_r$. Then each element $g \in F$ has a unique expression as a word $W(x_1, ..., x_r)$. Let $W_1, ..., W_k$ with $W_i = W_i(x_1, ..., x_r)$ be a set of words in the generators $x_1, ..., x_r$ of the free group F. At the most basic level, to construct a cryptosystem, suppose that we have a plaintext alphabet $\mathcal{A}$. For example suppose $\mathcal{A} = \{a, b, ...\}$ are the symbols needed to construct meaningful messages in English. To encrypt, use a substitution ciphertext

$$\mathcal{A} \to \{W_1, ..., W_k\}.$$

That is
$$a \longmapsto W_1, \ b \mapsto W_2, \cdots.$$

Then given a word $W(a, b, ...)$ in the plaintext alphabet form the free group word $W(W_1, W_2, ...)$ this represents an element $g \in F$. Send out g as the secret message. In order to implement this scheme we need a concrete representation of g and then for decryption a way to rewrite g back in terms of $W_1, ..., W_k$. This concrete representation is the idea behind **homomorphic cryptosystems** (see the article of Grigoriev and Ponomarenko [GP]).

The decryption algorithm then depends on the **Reidemeister-Schreier rewriting process** (see [MKS] for full details). Assume $W_1, ..., W_k$ are free generators for some subgroup H of F. A **Schreier transversal** for H is a set $\{h_1, ...h_t, ...\}$ of (left) coset representatives for H in F of a special form (again see [MKS] for particular details). Any subgroup of a free group has a Schreier transversal. The Reidemeister-Schreier process allows one to construct a set of generators $W_1, ..., W_k, ...$ for H by using a Schreier transversal. Further given the Schreier transversal from which the set of generators for H was constructed, the **Reidemeister-Schreier Rewriting Process** allows us to algorithmically rewrite an element of H. Given such an element expressed as a word $W = W(x_1, .., x_r)$ in the generators of F this algorithm rewrites W as a word $W^*(W_1, ..., W_k)$ in the generators of H. The actual algorithm is described in detail in [MKS] and [B]. What is important from the point of view of making codes secure is that given a set of generators it is easy to multiply them as strings in the whole free group. On the other hand even though the membership problem in a free group is solvable, that is it can determined if an element is in a subgroup, it is relatively difficult to determine a Schreier transversal. Although there are algorithms to accomplish this these are based for the most part on the knowledge of a homomorphic image of the subgroup and make it harder to determine a transversal given a set of generators than to rewrite using a known transversal. This is analogous to the situation for Grobner bases. Given a Grobner basis the explicit membership problem is computationally relatively easy. On the other hand it is computationally difficult to actually determine a Grobner basis. Work on the vulnerability of Grobner basis based schemes and word rewriting schemes in general can be found in the paper of Gonzalez-Vasco and Steinwandt [GS]. We note that rewriting in terms of the subgroup generators can also be accomplished in a free group via the Stalling foldings method and the construction of a subgroup graph (see [KM]).

In a general the free group F or general overgroup G is public. The subgroup H, where the encryption occurs, as well as the Schreier transversal, are kept secret. There are various attacks on the basic free group system (again see [GS]) so the method must be modified somewhat to thwart these. In [BFX 1,2] and [X] several ways to handle such attacks and a general schema for encrypting using free subgroups of finitely presented groups was discussed. In particular a random noise method to make them puiblically secure was introduced. This random noise method involved sending out as messages, not elements of the subgroup H, but rather elements in random right cosets of H (see [BFX 1,2]).

3. Some Proposed Platform Groups

Here we give some proposals for platform groups to be used with the group theoretic key transport protocol described in section 2. A group G is a candidate if it has either a nice finite presentation

$$G = \langle \ X; \ R \ \rangle$$

or a nice linear representation and has either a large abelian subgroup A or two large subgroups A_1, A_2 that commute elementwise. Although the word large here is ambiguous we mean large enough so that random choices can be made from them. In particular, for example, cyclic subgroups are inappropriate. There also should be some tie between the group used for the key exchange and the encryption method although this is not essential.

Our first proposed platform group G is $Aut(F_n)$, the automorphism group of a free group F_n of rank n. We assume that n is rather large, say at least 5, in order to get reasonable commuting subgroups. The secret key will be φ, a given automorphism of F_n. The encryption itself is in the free group F_n. We assume that we have a representation of F_n and an algorithm to go between the elements

of F_n and the representing objects. For example a linear representation and the representing matrices. As in the description of a free group cryptosystem given in the previous section we assume that we have a free subgroup $H = \langle W_1, ..., W_k \rangle$ and we encrypt by
$$a \longmapsto W_1, \ b \mapsto W_2, \cdots.$$
This free subgroup can be made public. As above, if our plain text message is $W(a, b, ...)$ written in the plaintext alphabet we form the free group element $W(W_1, W_2, ...)$. We then apply our secret automorphism φ to compute
$$\varphi(W) = W(\varphi(W_1), \varphi(W_2), ...)$$
in the free group F_n. We send out the represented form of this group element. If Alice is the decoder she rewrites the message in terms of the free group, computes φ^{-1} of the transmitted element and then has the appropriate form for decryption. An attacker must find the appropriate automorphism to steal the message.

$G = Aut(F_n)$ is an appropriate platfrom group for the proposed group theoretic Diffie-Hellman. There is a nice finite presentation for this group with the generators given by the Whitehead automorphisms (see [Mc 1,2] or [LS]). Further if the rank of the ambient free group F_n is large enough there is an abundant supply of large elementwise commuting subgroups in $Aut(F_n)$.

For example suppose $\{x_1, x_2, x_3, x_4, x_5, x_6\}$ is a free basis. Then those automorphisms which fix $\{x_1, x_2\}$ commute with those that fix $\{x_3, x_4, x_5\}$. This is the same idea exploited in braid group cryptography (see [AAG] or [D]).

Our second proposed platform group is $SL_4(\mathbb{Z})$. The secret key is a 4×4 unimodular integral matrix. For Bob's secret subgroup we take the subgroup isomorphic to $SL_2(\mathbb{Z})$ formed by considering those matrices that have elements of $SL_2(\mathbb{Z})$ in the upper left hand corner.

That is:
$$A_1 = \langle \begin{pmatrix} SL_2(\mathbb{Z}) & 0 \\ 0 & I \end{pmatrix} \rangle$$

For Alice's secret subgroup we take the corresponding subgroup formed by placing $SL_2(\mathbb{Z})$ isn the lower right hand corner
$$A_2 = \langle \begin{pmatrix} I & 0 \\ 0 & SL_2(\mathbb{Z}) \end{pmatrix} \rangle$$

To hide the special form of these subgroups Bob and Alice actually conjugate each of these subgroups by a fixed 4×4 unimodular integral matrix M. Actual encryption can be done in a free subgroup of $SL_4(\mathbb{Z})$ or with any other system. We note that there is a nice algorithm to determine free subgroups of $SL_4(\mathbb{Z})$ and free subgroups of $GL_n(\mathbb{Z})$ in general.

For $n \geq 3$ all the groups $GL_n(\mathbb{Z})$ are polycyclic-by-finite. In [BCRS] it was shown that there is an algorithm within polycyclic-by-finite groups that determines given a finite set of generators a presentation for the subgroup they generate. This algorithm can be applied in $GL_n(\mathbb{Z})$. From various results (see [C]) it follows that for a given m most subgroups of $GL_n(\mathbb{Z})$, generated by m randomly chosen matrices, are actually free with those matrices as a basis. Hence to generate a rank m free subgroup within $GL_n(\mathbb{Z})$, randomly choose m matrices in this group. Use the algorithm cited above to determine if these are a free basis. If yes we are done. If not randomly choose another set. From the probability considerations the probability is one that eventually a free subgroup will be chosen.

As a third possibility for a platform group we propose the surface braid groups. There are various kinds depending on the genus of the surface and whether it is orientable and/or punctured. The number of strands have to be considered too. The presentation of the surface group is finite and easy, the word problem is solvable. The pure braid group on a disc is a subgroup of the surface braid group, therefore its generators and pure braid relations appear in the presentation of the surface braid group. In addition to that we have generators that describe a strand going "through a wall." For a geometric interpretation we identify the surface group of genus g with a polygon with either $4g$ (orientable surface) or $2g$ (non-orientable surface) sides. A strand can go through the side of a polygon and either "reappear" from the opposite side (orientable case) or the adjacent side (non-orientable). We thus get $2g$ (g respectively) additional generators and many new relations. For details see [G] and [Be].

Like $SL_4(\mathbb{Z})$ the surface braid group also allows various systems for encryption. The surface braid group contains a free group and thus allows the use of the algorithm in free groups as mentioned above. Furthermore it contains commuting subgroups, not only the ones using the pure braid group generators but also ones where the additional generators occur. This provides more complexity for cryptographic computations than the pure braid group, For details see [C].

The final two proposed platforms are more theoretical in nature. The first is a free product of two free groups with commuting subgroups H, K. This has the presentational form

$$\langle F_1, F_1; [H, K] = 1 \rangle$$

where F_1 and F_2 are free groups and H and K are arbitrary subgroups of F_1 and F_2, respectively. For cryptographic purposes we assume that H and K are non-cyclic. Group theoretically such groups are free products with amalgamation (see [FRW]) and thus have nice normal forms. The commuting subgroups H and K can be used as the elementwise commuting subgroups for Bob and Alice. The structure of such groups was studied in [FRW] where a geometric Nielsen reduction theory was developed.

The final proposed platform group is a fully residually free group with a distinguished abelian subgroup B. This can be constructed in the following manner. Let F be a free group, $f \in F$ with f not a proper power. Then

$$G = \langle t, F;\ tft^{-1} = f \rangle$$

is called a rank one extension of centralizers of F. This is a fully residually free group (see [KhM] or [FMGRS] for terminology) and $\langle t, f \rangle$ generate a free abelian subgroup. We can iterate this procedure and keep track of the structure. We then obtain a fully residually free group with a large abelian subgroup. The theory of fully residually free groups, both structurally and algorithmically is well developed and played a role in the solution of the Tarski problem (see [KhM]). Because of the algorithmic similarities these groups seem to provide good possible extensions to free group cryptography.

We would like to thank the referee for many helpful comments as well as some very pertinent references.

4. References

[AN] R. Anderson and R. Needham, "Robustness Principle for Public Key Protocols," *Crypto 95*

[AAG] I.Anshel,M.Anshel and D.Goldfeld, "An Algebraic Method for Public Key Cryptography," *Math.Res. Lett* 6 (1999), 287-291, Springer Verlag

[B] G.Baumslag, *Topics in Combinatorial Group Theory* , Birkhauser, 1993

[Be] P.Bellingeri, "On Presentations of Surface Braid Groups," *J. Algebra*, 274 (2), (2004), 543-563

[BMS] A.Borovik,A.Myasnikov and V. Shpilrain, "Measuring Sets in Infinite Groups," *Cont. Math.*, 298, (2002), 21-42

[BM] C.Boyd and A. Mathuria, *Protocols for Authentication and Key Establishment,* Springer-Verlag, 2003

[BFX 1] G.Baumslag, B.Fine,and X.Xu, "Cryptosystema Using the Linear Groups," *Proceeding of Algebraic Cryptography,* (2005)

[BFX 2] G.Baumslag, B.Fine,and X.Xu, "A Proposed Public Key Cryptosystem Using the Modular Group," *Cont. Math* , (2006), to appear

[BCRS] G.Baumslag,F.Cannonito,D.Robinson,D.Segal, "The Algorithmic Theory of Polycyclic-by-Finite Groups ," *J. Alg.* , 142 , (1991), 118-149

[BG] AS.R. Blackburn and S. Galbraith, "Cryptanalysis of Two Cryptosystems Nased on Group Actions," *Advances in Cryptology ASIACRYPT 99,* Lecture Notes in Computer Science, 1716 , (1999), 52-61

[C] T.Camps, "Cryptography Using the Surface Braid Groups," thesis University of Dortmund, In Progress

[D] P.Dehornoy, "Braid Based Cryptography ," *Cont. Math.* , 360 ,(2004), 5-34

[F] B.Fine, *The Algebraic Theory of the Bianchi Groups,* Marcel Dekker, 1990

[FGMRS] B.Fine,A.Gaglione,A.Myasnikov,G.Rosenberger and D.Spellman, "A Classification of Fully Residually Free Groups ," *J. of Algebra* , 200, (1998), 571-605

[FRW] B.Fine,G.Rosenberger and R.Weidmann, "Two Generator Subgroups of Free Products with Commuting Subgroups ," *J.Pure and Applied Algebra* , 172, (2002), 193-204

[G] PJ.Gonzalez-Meneses, "New Presentations of Surface Braid Groups," *J. Knot Theory Algebra*, 10 (3), (2001), 431-451

[GS] M.I.Gonzalez-Vasco and R. Steinwandt, "Chosen Ciphertext Attacks as Common Vulnerability of Some Group and Polynomial Based Encryption Schemes," *Tatra Mountains Mathematical Publications,* (2005), to appear

[GP] D.Grigoriev and I. Ponomarenko, "Homomorphic Public-Key Cryptosystems Over Groups and Rings ," *Quaderni di Matematica,* (2005), to appear

[HGS] C.Hall, I. Goldberg, B. Schneier, "Reaction attacks Against Several Public Key Cryptosystems," *Proceedings of Information and Communications Security ICICS 99,* Springer-Verlag (1999), 2-12

[KM] I. Kapovich and A. Myasnikov, "Stallings Foldings and Subgroups of Free Groups," *J. Algebra*, 248, (2002), 608-668

[KhM] O.Kharlampovich and A. Myasnikov, *Algebraic Geometry over Free Groups,* to appear

[KLCHKP] K.H.Ko, S.J. Lee, J.H. Cheon,J.W. Han, J.Kang and C.Park, "New Public Key Cryptosystem Using Braid Groups," *Advances in Cryptology CRYPTO 2000,* Lecture Notes in Computer Science, 1880, (2000), 166-183

[Ko] N.Koblitz, *Algebraic Methods of Cryptography,* Springer , 1998

[LS] R.Lyndon and P.Scupp, *Combinatorial Group Theory,* Springer , 1978

[MKS] W. Magnus, A. Karass and D. Solitar, *Combinatorial Group Theory,* Wiley Interscience, New York , 1968

[Mc 1] J. McCool, "A Presentation for the Automorphism Group of a Free Group of Finite Rank ," *J. London Math. Soc.* , 8 , (1974), 259-266

[Mc 2] J. McCool, "On Nielsen's Presentation for the Automorphism Group of a Free Group ," *J. London Math. Soc.* , 10 , (1975), 265-270

[PHKCP] S.H.Paeng, K.C.Ha,J.H.Kim,S.Chee and C. Park, "New Public Key Cryptosystem Using Finite NonAbelian Groups," *Advances in Cryptology - Crypto 2001,* Springer, (2001), 470-483

[SU] V.Shpilrain and A. Ushakov, "Thompson's Group and Public Key Cryptography," *ACNS 2005 Lecture Notes in Computer Science,* 3531 , (2005), 151-164

[S] R. Steinwandt, "Loopholes in two public key cryptosystems using the modular groups," preprint Univ. of Karlsruhle, (2000)

[X] Xiaowei Xu, "Cryptography and Infinite Group Theory Ph.D. Thesis CUNY, 2005

[Y] A.Yamamura, "Public Key cryptosystems using the modular groups," *Lecture Notes in Comput. Sci.* 1431 , (1998), 203-216

DEPARTMENT OF MATHEMATICS, CITY COLLEGE OF NEW YORK, NEW YORK, NY 10031

FACHBEREICH MATHEMATIK, UNIVERSITAT DORTMUND, 44227 DORTMUND, FEDERAL REPUBLIC OF GERMANY

DEPARTMENT OF MATHEMATICS, FAIRFIELD UNIVERSITY, FAIRFIELD, CT 06430

FACHBEREICH MATHEMATIK, UNIVERSITAT DORTMUND, 44227 DORTMUND, FEDERAL REPUBLIC OF GERMANY

DEPARTMENT OF MATHEMATICS, CITY COLLEGE OF NEW YORK, NEW YORK, NY 10031

Geometric Key Establishment

Arkady Berenstein and Leon Chernyak

ABSTRACT. We propose a new class of key establishment schemes which are based on geometric generalizations of the classical Diffie-Hellman. The simplest of our schemes – based on the geometry of the unit circle – uses only multiplication of rational numbers by integers and addition of rational numbers in its key creation. Its first computer implementation works significantly faster than all known implementations of Diffie-Hellman. Preliminary estimations show that our schemes are resistant to attacks. This resistance follows the pattern of the discrete logarithm problem and hardness of multidimensional lattice problems.

Contents

0. Introduction
1. Key establishment schemes based on commuting double actions
2. Geometric Key Establishment I
3. Geometric Key Establishment II
4. Appendix A. General Key Establishment Scheme and its rounded versions
5. Appendix B. Proof of results of Sections 2 and 3

References

0. Introduction

In this paper we propose a new class of key establishment schemes which we refer to as Geometric Key Establishment (GKE). Similarly to Diffie-Hellman ([5]), the GKE schemes do not assume that communicating parties share any kind of secret information prior to the act of key creation and distribution.

The GKE schemes are based on the mathematical concept of semigroup action and its modification – commuting double action. Cryptographic applications of the semigroup actions are well-known: Diffie-Hellman schemes are based on actions of the semigroup of integers (under multiplication) on finite groups. More recent applications include two-sided multiplications in semigroups and groups ([8]), actions of semigroups of square integer matrices on finite commutative groups ([7]) and actions of braid semigroups on braid groups ([2]).[1]

Although general commuting double actions seem to be well-known in mathematics, we are unaware of any application of this concept in key establishment protocols.[2] In the present work we construct two geometric key establishment schemes (GKE I and GKE II) which are based primarily on the concept.

Typically, Diffie-Hellman-like schemes involve time-consuming exponentiation procedures in finite fields or finite groups. Unlike this, GKE I and GKE II do not use any exponentiation. We bypassed exponentiation by replacing the semigroup actions on finite groups with actions (or commuting double actions) on infinite and even continuous groups. In particular, the simplest of our schemes is based on the action of the semigroup of integer square matrices on the unit cube, and, by design, uses only multiplication of real numbers by integers and addition of real numbers in its key creation.

First computer implementations of the cube-based schemes work with a much higher speed than all known implementations of Diffie-Hellman. More precisely, the running time of GKE I (resp. GKE II) is proportional to N^2 (resp. $N^{3/2}$) with the assumption that multiplication of two numbers takes a constant time, where N is the size of the input.

Preliminary estimations show that GKE I and GKE II are resistant to basic attacks. This resistance follows the pattern of the discrete logarithm problem. A more detailed study of GKE security is a work [4] joint with Professor Itkis of Boston University.

As we said above, our schemes rely on infinite geometric objects or, more precisely, on compact connected topological groups such as the unit circle or an n-dimensional torus. Of course, the schemes, as based on infinite geometric objects, are *ideal* in that sense that no *real* computing device can create or communicate keys as points of a geometric continuum. In order to implement GKE in a real device, we developed (based on the ordinary rounding of real numbers) a procedure of *discretization* of our ideal, continuous schemes. This procedure allows for creating an infinite family of *real* key establishment protocols. These real protocols seem to be cryptographically sound, which fact is by itself very inspiring.

Having been encouraged by obtaining a rich family of discretizations for GKE, we proceeded to generalization of the relationship between ideal and real key establishment schemes. As a result, we introduced a general concept of Rounded Key Establishment (RKE). This latter concept consists of an ideal continuous scheme and a family of its discretizations. One of surprising results of this generalization is a rigorous mathematical definition of key establishment, in which all existing Diffie-Hellman-like schemes fit perfectly. We have not been able to find any reference to similarly rigorous mathematical definition of key establishment in the literature.

We hope that, in addition to GKE I and GKE II, our concept of RKE will bring new interesting examples of key establishment schemes.

The paper is organized as follows:

---

[1] An approach not based on Diffie-Hellman has been suggested by I. Anshel, M. Anshel, and D. Goldfeld in [1] and [3].

[2] The approach developed in [8] utilizes a particular case of two-sided action. The limitation of this approach consists in the requirement that the involved semigroups are commutative.

In *Section* 1 we introduce key establishment paradigms based on commuting double actions. Our main examples include all schemes based on semigroup and ring actions and, in particular, Diffie-Hellman scheme and its generalizations. Our examples will be used in the following sections for constructing our GKE I and GKE II schemes.

Section 2 is devoted to introduction and study of our first main example – Geometric Key Establishment I (GKE I). We start with a description of an ideal GKE I and then construct a family of its discretizations. The main result of the section is Theorem 2.2, which asserts that these discretizations bring about a family of real key establishment protocols. We conclude the section with a numerical example demonstrating how the real GKE I protocols work.

Section 3 is devoted to introduction and study of our second main example –Geometric Key Establishment II (GKE II). The section is structured similarly to Section 2. We start with a description of an ideal GKE II and then construct a family of its discretizations. The main result of the section is Theorem 3.2, which asserts that these discretizations bring about a family of real key establishment protocols. We conclude the section with a numerical example demonstrating how the real GKE II protocols work.

In *Appendix A* we develop a conceptual framework for rounded key establishment (RKE). The basic key establishment scheme (Definition A.1) is quite trivial and, apparently, is well known (although we have been unable to find appropriate references). However, having been written in the set-theoretic language, it allows for a simple conceptual definition of RKE. This approach is common in modern mathematics: once an object is defined set-theoretically, it can further be enriched topologically, algebraically, and geometrically.

Appendix B consists of the proofs of main results – Theorem 2.2 and Theorem 3.2.

Acknowledgments. The authors express their gratitude to Igor Mendelev for invaluable help in implementation of the first prototype of GKE and for performing the comparative analysis of GKE prototype with other key establishment systems. Our thanks are due to Professor Itkis of Boston University for extremely helpful comments and remarks on this manuscript. The authors wish to thank Professor Michael Anshel of City College of New York for very helpful references in the field of the semigroup-based cryptography. We would like to express our gratitude to Professor Shpilrain for giving us the opportunity to present this work at the "Algebraic Cryptography" section of Canadian Mathematical Society conference held in December 2004.

1. Key establishment schemes based on commuting double actions

In this section we introduce a class of key establishment schemes which we refer to as *commuting double action* schemes. This class of schemes is based on the mathematical concept of *commuting double action*.

DEFINITION 1.1. Let A, B, and X be sets. A quadruple of maps $A \times X \to X$, $X \times A \to X$ and $B \times X \to X$, $X \times B \to X$ (denoted respectively as: $(\mathbf{a}, \mathbf{x}) \to \mathbf{a}(\mathbf{x})$, $(\mathbf{x}, \mathbf{a}) \to (\mathbf{x})\mathbf{a}$ and $(\mathbf{b}, \mathbf{x}) \to \mathbf{b}(\mathbf{x})$, $(\mathbf{x}, \mathbf{b}) \to (\mathbf{x})\mathbf{b}$ is a *commuting double action* of A and B on X if:

$$(1.1) \qquad (\mathbf{a}(\mathbf{x}))\mathbf{b} = (\mathbf{b}(\mathbf{x}))\mathbf{a}$$

for any $\mathbf{x} \in X$ and any $\mathbf{a} \in A$, $\mathbf{b} \in B$.

Commuting double action key establishment scheme

Setup (non-secret parameters)
- Sets A, B, and X
- a commuting double action $A \times X \to X$, $X \times A \to X$ and $B \times X \to X$, $X \times B \to X$

Protocol
- *Alice* and *Bob*: choose a non-secret element $\mathbf{x} \in X$
- *Alice*: choose a secret element $\mathbf{a} \in A$
- *Bob*: choose a secret element $\mathbf{b} \in B$
- *Alice→Bob*: $m_A = \mathbf{a}(\mathbf{x})$
- *Bob*: compute $S_B = (m_A)\mathbf{b} = (\mathbf{a}(\mathbf{x}))\mathbf{b}$
- *Bob→Alice*: $m_B = \mathbf{b}(\mathbf{x})$
- *Alice*: compute $S_A = (m_B)\mathbf{a} = (\mathbf{b}(\mathbf{x}))\mathbf{a}$

Common Secret

By Definition 1.1, $S_A = S_B$.

REMARK 1.2. If the m_A, S_A, m_B, S_B are computed with some precision, then one has $S_A \approx S_B$ and additional exchange between Alice and Bob may be necessary.

Now we consider examples of commuting double actions coming from semigroups and their actions on sets.

DEFINITION 1.3. A *semigroup* is a set A with an associative multiplication $A \times A \to A$, i.e.
$$(\mathbf{ab})\mathbf{c} = \mathbf{a}(\mathbf{bc}) \ .$$
for any $\mathbf{a}, \mathbf{b}, \mathbf{c} \in A$.

EXAMPLE 1.4. Let X be a semigroup (i.e., a set with an associative multiplication) and let $A \subseteq X$ and $B \subseteq X$ be any subsets. Then the maps $A \times X \to X$, $X \times A \to X$ and $B \times X \to X$, $X \times B \to X$ given respectively by:
$$\mathbf{a}(\mathbf{x}) = (\mathbf{x})\mathbf{a} = \mathbf{a} \cdot \mathbf{x} \text{ and } \mathbf{b}(\mathbf{x}) = (\mathbf{x})\mathbf{b} = \mathbf{x} \cdot \mathbf{b}$$
constitute a commuting double action because
$$(\mathbf{a}(\mathbf{x}))\mathbf{b} = (\mathbf{a} \cdot \mathbf{x}) \cdot \mathbf{b} = \mathbf{a} \cdot (\mathbf{x} \cdot \mathbf{b}) = (\mathbf{b}(\mathbf{x}))\mathbf{a} \ .$$

REMARK 1.5. A slightly more general example of two-sided multiplication in semigroups and groups was considered in [8].

DEFINITION 1.6. Let M be a semigroup and let X be a set. A *left* action of M on X is a map $M \times X \to X$ (to be denoted by $(\mathbf{a}, \mathbf{x}) \to \mathbf{a} \cdot \mathbf{x}$ for any $\mathbf{a} \in A$, $\mathbf{x} \in X$) such that
$$(1.2) \qquad \mathbf{a}(\mathbf{bx}) = (\mathbf{a} \cdot \mathbf{b})\mathbf{x}$$
for any elements $\mathbf{a}$ and $\mathbf{b}$ of M and any $\mathbf{x} \in X$. A *right* action of a semigroup M on X is a map $M \times X \to X$ (to be denoted by $(\mathbf{x}, \mathbf{a}) \to \mathbf{xa}$ for any $\mathbf{a} \in M$, $\mathbf{x} \in X$) such that
$$(1.3) \qquad (\mathbf{xa})\mathbf{b} = \mathbf{x}(\mathbf{a} \cdot \mathbf{b})$$
for any elements $\mathbf{a}$ and $\mathbf{b}$ of M and any $\mathbf{x} \in X$.

DEFINITION 1.7. Let M be a semigroup. Given two sets A and B and two pairs of maps $\varphi_A, \varphi'_A : A \to M$ and $\varphi_B, \varphi'_B : B \to M$, we say that the quadruple $(\varphi_A, \varphi'_A, \varphi_B, \varphi'_B)$ *quasi-commutes* if

$$\varphi'_B(\mathbf{b}) \cdot \varphi_A(\mathbf{a}) = \varphi'_A(\mathbf{a}) \cdot \varphi_B(\mathbf{b}) \tag{1.4}$$

for all $\mathbf{a} \in A$ and $\mathbf{b} \in B$.

LEMMA 1.8. *Let M be a semigroup and let X, A, and B be sets. Fix a quasi-commuting quadruple of maps $\varphi_A, \varphi'_A : A \to M$ and $\varphi_B, \varphi'_B : B \to M$. Then:*
(a) For any left action $M \times X \to X$ the following four maps $A \times X \to X$, $X \times A \to X$ and $B \times X \to X$, $X \times B \to X$ constitute a commuting double action of A and B on X:

$$\mathbf{a}(\mathbf{x}) = \varphi_A(\mathbf{a})\mathbf{x}, \mathbf{b}(\mathbf{x}) = \varphi_B(\mathbf{b})\mathbf{x},\ \mathbf{x})\mathbf{a} = \varphi'_A(\mathbf{a})\mathbf{x},\ (\mathbf{x})\mathbf{b} = \varphi'_B(\mathbf{b})\mathbf{x} \tag{1.5}$$

for any $\mathbf{x} \in X$ and any $\mathbf{a} \in A$, $\mathbf{b} \in B$.
(b) For any right action $X \times M \to X$ the following four maps $A \times X \to X$, $X \times A \to X$ and $B \times X \to X$, $X \times B \to X$ constitute a commuting double action of A and B on X:

$$\mathbf{a}(\mathbf{x}) = \mathbf{x}\varphi'_A(\mathbf{a}),\ \mathbf{b}(\mathbf{x}) = \mathbf{x}\varphi'_B(\mathbf{b}),\ (\mathbf{x})\mathbf{a} = \mathbf{x}\varphi_A(\mathbf{a}),\ (\mathbf{x})\mathbf{b} = \mathbf{x}\varphi_B(\mathbf{b}) \tag{1.6}$$

for any $\mathbf{x} \in X$ and any $\mathbf{a} \in A$, $\mathbf{b} \in B$.

Proof. Prove (a). For any $\mathbf{x} \in X$ and any $\mathbf{a} \in A$, $\mathbf{b} \in B$ we have:

$$(\mathbf{a}(\mathbf{x}))\mathbf{b} = \varphi'_B(\mathbf{b})(\varphi_A(\mathbf{a})\mathbf{x}) = (\varphi'_B(\mathbf{b}) \cdot \varphi_A(\mathbf{a}))\mathbf{x} =$$
$$= (\varphi'_A(\mathbf{a}) \cdot \varphi_B(\mathbf{b}))\mathbf{x} = \varphi'_A(\mathbf{a})(\varphi_B(\mathbf{b}))\mathbf{x}) = (\mathbf{b}(\mathbf{x}))\mathbf{a} \ .$$

This proves (a). Prove (b) now.

$$(\mathbf{a}(\mathbf{x}))\mathbf{b} = (\mathbf{x}\varphi'_A(\mathbf{a}))\varphi_B(\mathbf{b}) = \mathbf{x}(\varphi'_A(\mathbf{a}) \cdot \varphi_B(\mathbf{b})) =$$
$$= \mathbf{x}(\varphi'_B(\mathbf{b}) \cdot \varphi_A(\mathbf{a})) = (\mathbf{x}\varphi'_B(\mathbf{b}))\varphi_A(\mathbf{a}) = (\mathbf{b}(\mathbf{x}))\mathbf{a} \ .$$

This proves the lemma. □

DEFINITION 1.9. Let M be a semigroup and let (α, α') and (β, β') be two pairs of elements of M. We say that these two pairs quasi-commute if

$$\beta' \cdot \alpha = \alpha' \cdot \beta \ . \tag{1.7}$$

More generally, given two families

$$A = \{(\alpha_i, \alpha'_i) | i = 1, 2, \ldots, k\}, B = \{(\beta_j, \beta'_j) | j = 1, 2, \ldots, l\}$$

of pairs of elements of M, we say that A and B *quasi-commute* if

$$\beta'_j \cdot \alpha_i = \alpha'_i \cdot \beta_j \tag{1.8}$$

for all $i = 1, 2, \ldots, k$, $j = 1, 2, \ldots, l$.

The following example links Definitions 1.7 and 1.9 in the case when M is a ring.

EXAMPLE 1.10. Let M be a ring and let
$$A = \{(\alpha_i, \alpha_i') | i = 1, 2, \ldots, k\}, B = \{(\beta_j, \beta_j') | j = 1, 2, \ldots, l\}$$
be quasi-commuting families of pairs of elements in M. Define four maps $\varphi_A, \varphi_A' : \mathbb{Z}^k \to M$ and $\varphi_B, \varphi_B' : \mathbb{Z}^l \to M$ by the formula:

$$(1.9) \quad \varphi_A(\mathbf{a}) = \sum_{i=1}^k a_i \cdot \alpha_i, \varphi_A'(\mathbf{a}) = \sum_{i=1}^k a_i \cdot \alpha_i', \varphi_B(\mathbf{b}) = \sum_{j=1}^l b_j \cdot \beta_j, \varphi_B'(\mathbf{b}) = \sum_{j=1}^l b_j \cdot \beta_j'$$

for any $\mathbf{a} = (a_1, a_2, \ldots, a_k) \in \mathbb{Z}^k$, $\mathbf{b} = (b_1, b_2, \ldots, b_l) \in \mathbb{Z}^l$.

LEMMA 1.11. *Let M be a ring and let*
$$A = \{(\alpha_i, \alpha_i') | i = 1, 2, \ldots, k\}, B = \{(\beta_j, \beta_j') j = 1, 2, \ldots, l\}$$
be two quasi-commuting families of elements of M. Then the quadruple of maps $\varphi_A, \varphi_A' : \mathbb{Z}^k \to M$ and $\varphi_B, \varphi_B' : \mathbb{Z}^l \to M$ defined by (1.9) quasi-commutes (in the sense of Definition 1.7).

Proof. Define the map $\Psi : \mathbb{Z}^k \times \mathbb{Z}^l \to M$ by the formula
$$\Psi(a, b) = \varphi_B'(\mathbf{b}) \cdot \varphi_A(\mathbf{a}) - \varphi_A'(\mathbf{a}) \cdot \varphi_B(\mathbf{b})$$
for any $\mathbf{a} \in \mathbb{Z}^k$, $\mathbf{b} \in \mathbb{Z}^l$. In view of (1.4), our goal is to prove that $\Psi = 0$. Indeed, Ψ is linear in both $\mathbf{a}$ and $\mathbf{b}$, that is:
$$\Psi(a, b) = \sum_{i=1}^k \sum_{j=1}^l a_i b_j \cdot \Psi(e_i, f_j).$$

where $e_1, e_2, \ldots, e_k$ is the standard basis for $\mathbb{Z}^k$ and $f_1, f_2, \ldots, f_l$ is the standard basis for $\mathbb{Z}^l$. On the other hand, by definition, $\Psi(e_i, f_j) = \beta_j' \cdot \alpha_i - \alpha_i' \cdot \beta_j = 0$ for all i and j by (1.7). Therefore, $\Psi = 0$. The lemma is proved. $\square$

REMARK 1.12. A key establishment scheme utilizing a particular case of Example 1.10 (in the case when all α_i, α_i', β_j, β_j' are certain powers of an element $S \in M$ was suggested in [7].

EXAMPLE 1.13. (Diffie-Hellman). In the notation of Definition 1.4 let $M = A = B = \mathbb{Z}$, the set of all integers considered a semigroup under multiplication. Then raising elements of any group X into integer powers defines a commuting double action of M on X via:
$$\mathbf{a(x)} = \mathbf{(x)a} = \mathbf{x}^a, \ \mathbf{b(x)} = \mathbf{(x)b} = \mathbf{x}^b$$
because $(\mathbf{x}^a)^b = \mathbf{x}^{ab}$.

In what follows we denote by $M_n(\mathbb{Z})$ the set of integer $n \times n$ matrices. This is a ring under matrix addition and matrix multiplication.

Our first main example below will generalize all examples of semigroup actions constructed in [7].

Main Example I
- G is any group
- n is any natural number
- G^n denotes the n-th Cartesian power of G, i.e., the set of all n-tuples $\mathbf{g} = (g_1, \ldots, g_n)$ of elements of G

- $X_n = [G^n] \subseteq G^n$ is the set of all pairwise commuting tuples $\mathbf{g} = (g_1, \ldots, g_n)$, i.e.,

$$g_i \cdot g_j = g_j \cdot g_i$$

for $i, j = 1, 2, \ldots, n$

- Two families of quasi-commuting integer $n \times n$ matrices

$$A = \{(\alpha_i, \alpha'_i) | i = 1, 2, \ldots, k\}, B = \{(\beta_j, \beta'_j) | j = 1, 2, \ldots, l\}$$

- A map (not action!) $G^n \times M_n(\mathbb{Z}) \to G^n$ is given by the formula: $(\mathbf{g}, \mathbf{A}) \to \mathbf{g}^\mathbf{A}$

for any $\mathbf{A} = (a_{ij}) \in M_n(\mathbb{Z})$, $\mathbf{g} = (g_1, \ldots, g_n) \in [G^n]$, where $\mathbf{g}^A$ is the $\mathbf{A}$-th power of $\mathbf{g}$:

$$\mathbf{g}^\mathbf{A} = (g'_1, \ldots, g'_n),$$

where

(1.10) $$g'_j = \prod_{i=1}^n g_i^{a_{ij}}$$

LEMMA 1.14. *The assignment* $(\mathbf{g}, \mathbf{A}) \to \mathbf{g}^\mathbf{A}$ *is a right action of* $M_n(\mathbb{Z})$ *on* $X_n = [G^n]$:

(1.11) $$X_n \times M_n(\mathbb{Z}) \to X_n,$$

i.e., for any $\mathbf{A} = (a_{ij})$, $\mathbf{B} = (b_{ij}) \in M_n(\mathbb{Z})$ *and any* $\mathbf{g} \in X_n$ *one has*

(1.12) $$(\mathbf{g}^A)^B = \mathbf{g}^{AB}.$$

Proof. It suffices to prove only (1.12). Indeed, using the fact that all g_i commute with each other, we have by (1.10) for all j:

$$((\mathbf{g}^A)^B)_j = \prod_{k=1}^n (\mathbf{g}^A)_k^{b_{kj}} = \prod_{k=1}^n \left(\prod_{i=1}^n g_i^{a_{ik}} \right)^{b_{kj}} = \prod_{i=1}^n g_i^{c_{ij}},$$

where $c_{ij} = \sum_{k=1}^n a_{ik} b_{kj} = (\mathbf{AB})_{ij}$.

Therefore, $((\mathbf{g}^A)^B)_j = (\mathbf{g}^{AB})_j$ for all j. This proves (1.12). The lemma is proved. $\square$

Below we propose our second main example of a commuting double action.

Main Example II
- G is a group.
- m and n are natural numbers
- $M_{m \times n}(G)$ denotes the set of $m \times n$ matrices $\mathbf{g} = (g_{ij})$ with coefficients $g_{ij} \in G$
- $X_{m \times n} = [M_{m \times n}(G)] \subseteq M_{m \times n}(G)$ is the set of all those elements $\mathbf{g} = (g_{ij}) \in M_{m \times n}(G)$ in which the entries pairwise commute, i.e.,

$$g_{ij} \cdot g_{kl} = g_{kl} \cdot g_{ij}$$

for all $i, k = 1, 2, \ldots, m; j, l = 1, 2, \ldots, n$
- Maps (not a commuting double action!)

$$M_m(\mathbb{Z}) \times M_{m \times n}(G) \to M_{m \times n}(G), \ M_{m \times n}(G) \times M_m(\mathbb{Z}) \to M_{m \times n}(G)$$
$$M_n(\mathbb{Z}) \times M_{m \times n}(G) \to M_{m \times n}(G), \ M_{m \times n}(G) \times M_n(\mathbb{Z}) \to M_{m \times n}(G)$$

given by $(\mathbf{g})\mathbf{A} = \mathbf{A}(\mathbf{g}) = {}^{\mathbf{A}}\mathbf{g}, (\mathbf{g})\mathbf{B} = \mathbf{B}(\mathbf{g}) = \mathbf{g}^{\mathbf{B}}$, where ${}^{\mathbf{A}}\mathbf{g} = (g'_{ij})$, $\mathbf{g}^{\mathbf{B}} = (g''_{ij})$ are given by the formula:

$$g'_{ij} = \prod_{k=1}^{n} g_{kj}^{a_{ik}}, \ g''_{ij} = \prod_{k=1}^{n} g_{ik}^{b_{kj}}.$$

LEMMA 1.15. *These data define a commuting double action* $M_m(\mathbb{Z}) \times X_{m \times n} \to X_{m \times n}$ *and* $X_{m \times n} \times M_n(\mathbb{Z}) \to X$, *i.e.,* $({}^{\mathbf{A}}\mathbf{g})^{\mathbf{B}} = {}^{\mathbf{A}}(\mathbf{g}^{\mathbf{B}})$ *for any* $\mathbf{A} \in M_m(\mathbb{Z})$, $\mathbf{B} \in M_n(\mathbb{Z})$, $\mathbf{g} \in X_{m \times n}$.

Proof. It is equivalent to the associativity of the matrix multiplication:

$$(\mathbf{Ax})\mathbf{B} = \mathbf{A}(\mathbf{xB})$$

for any $m \times m$ matrix $\mathbf{A}$, any $m \times n$ matrix $\mathbf{x}$, and $n \times n$ matrix $\mathbf{B}$. The lemma is proved. □

2. Geometric Key Establishment I

In this section we present a key establishment scheme based on **Main Example I** and on the general right action key establishment scheme of Section 1. We will refer to it as geometric key establishment I (GKE I). First, we present the ideal GKE scheme (i.e., without any rounding).

Ideal Geometric Key Establishment I (GKE I) Scheme

Setup (non-secret parameters)

- n is a natural number
- $X_n = [0,1)^n$ is the semi-open n-dimensional cube, i.e., X_n is the n-th Cartesian power of the semi-open interval $[0,1)$ of the real line. A point of X_n is an n-tuple $\mathbf{g} = (g_1, g_2, \ldots, g_n)$, where each $g_i \in [0,1)$
- Two families of quasi-commuting integer $n \times n$ matrices

  $$A = \{(\alpha_i, \alpha'_i) | i = 1, 2, \ldots, k\}, B = \{(\beta_j, \beta'_j) | j = 1, 2, \ldots, l\}$$

- The right action $X_n \times M_n(\mathbb{Z}) \to X_n$ is given by the formula

(2.1) $$(\mathbf{g}, \mathbf{A}) \to \mathbf{g} \cdot \mathbf{A}$$

  for any matrix $\mathbf{A} \in M_n(\mathbb{Z})$ and any $\mathbf{g} \in \mathbf{X}_n$, where for each vector $\mathbf{x} = (x_1, x_2, \ldots, x_n)$ of real numbers we use the notation $\mathbf{x} = (\{x_1\}, \{x_2\}, \ldots, \{x_n\})$, where $\{z\}$ stands for the fractional part of a real number z.

LEMMA 2.1. *For any* $\mathbf{a} = (a_1, a_2, \ldots, a_k) \in \mathbb{Z}^k$, $\mathbf{b} = (b_1, b_2, \ldots, b_l) \in \mathbb{Z}^l$, *and* $\mathbf{g} \in X_n$ *one has*

$$\{\{\mathbf{g} \cdot \varphi_A(\mathbf{a})\} \cdot \varphi'_B(\mathbf{b})\} = \{\mathbf{g} \cdot \varphi_A(\mathbf{a}) \cdot \varphi'_B(\mathbf{b})\} = \{\mathbf{g} \cdot \varphi_B(\mathbf{b}) \cdot \varphi'_A(\mathbf{a})\} = \{\{\mathbf{g} \cdot \varphi_B(\mathbf{b})\} \cdot \varphi'_A(\mathbf{a})\},$$

where $\varphi_A(\mathbf{a}), \varphi'_A(\mathbf{a}), \varphi_B(\mathbf{b})$, *and* $\varphi'_B(\mathbf{b})$ *are defined in* (1.9).

Proof. Taking the group $G = [0,1)$ with the operation $\alpha * \beta = \{\alpha + \beta\}$ in Lemma 1.14 we obtain

$$\{\{\mathbf{g} \cdot \mathbf{A}\} \cdot \mathbf{B}\} = \{\mathbf{g} \cdot \mathbf{A} \cdot \mathbf{B}\}$$

for any $\mathbf{g} \in X_n$ and any $\mathbf{A}, \mathbf{B} \in M_n(\mathbb{Z})$.

This and the quasi-commutation equation (1.4) prove the lemma. □

Protocol
- *Alice* and *Bob*: choose a non-secret $\mathbf{g} \in X_n$
- *Alice*: choose a secret $\mathbf{a} = (a_1, a_2, \ldots, a_k) \in \mathbb{Z}^k$
- *Bob*: choose a secret $\mathbf{b} = (b_1, b_2, \ldots, b_l) \in \mathbb{Z}^l$
- *Alice*→*Bob*: $m_A = \{\mathbf{g} \cdot \varphi_A(\mathbf{a})\}$
- *Bob*: compute $S_B = \{m_A \cdot \varphi'_B(\mathbf{b})\} = \{\{\mathbf{g} \cdot \varphi_A(\mathbf{a})\} \cdot \varphi'_B(\mathbf{b})\}$
- *Bob*→*Alice*: $m_B = \{\mathbf{g} \cdot \varphi_B(\mathbf{b})\}$
- *Alice*: compute $S_A = \{m_B \cdot \varphi'_A(\mathbf{a})\} = \{\{\mathbf{g} \cdot \varphi_B(\mathbf{b})\} \cdot \varphi'_A(\mathbf{a})\}$

Common Secret

By Lemma 2.1, $S_A = S_B$.

In order to present the *rounded* GKE I, we need the following notation. For any real vectors $y = (y_1, y_2, \ldots, y_n)$, $z = (z_1, z_2, \ldots, z_n)$ the vector inequality $y \leq z$ is equivalent to n scalar inequalities:

$$y_1 \leq z_1, \ y_2 \leq z_2, \ldots, y_n \leq z_n .$$

Also the inequality $|y| < z$ means that $y < z$ and $-y < z$.

Denote by $Round(z)$ the standard rounding of a real number z to the closest integer. Also for any real number $g \in [0, 1)$ and any natural number P denote:

(2.2) $$[g]_P = \begin{cases} (Round(gP))/P & \text{if } Round(gP) < P \\ [g]_P = 0 & \text{if } Round(gP) = P \end{cases}.$$

For any natural n-tuple $\mathbf{P} = (P_1, P_2, \ldots, P_n)$ and a real n-vector $\mathbf{g} = (g_1, g_2, \ldots, g_n)$ such that each $g_i \in [0, 1)$, we define the $\mathbf{P}$-rounding to a rational n-tuple $[\mathbf{g}]_{\mathbf{P}}$ by:

$$[\mathbf{g}]_{\mathbf{P}} = ([g_1]_{P_1}, [g_2]_{P_2}, \ldots, [g_n]_{P_n}) .$$

For an n-tuple $\mathbf{P} = (P_1, P_2, \ldots, P_n)$ of natural numbers denote $\mathbf{P}^* = (\frac{1}{P_1}, \frac{1}{P_2}, \ldots, \frac{1}{P_n})$.

THEOREM 2.2. *Let $\mathbf{P}, \mathbf{Q}$, and $\mathbf{K}$ be natural n-tuples. Then for any real n-vector $\mathbf{g}$ and any integer $n \times n$ matrices $\mathbf{A} = (a_{ij})$, $\mathbf{B} = (b_{ij})$, $\mathbf{A}' = (a'_{ij})$ and $\mathbf{B}' = (b'_{ij})$ satisfying $\mathbf{A} \cdot \mathbf{B}' = \mathbf{B} \cdot \mathbf{A}'$ and $\mathbf{Q}^* \cdot |\mathbf{A}'| \leq \mathbf{K}^*$, $\mathbf{P}^* \cdot |\mathbf{B}'| \leq \mathbf{K}^*$ one has: either at least one coordinate of $[\{[\{\mathbf{g} \cdot \mathbf{A}\}]_{\mathbf{P}} \cdot \mathbf{B}'\}]_{\mathbf{K}}$ equals 0, or at least one coordinate of $[\{[\{\mathbf{g} \cdot \mathbf{B}\}]_{\mathbf{Q}} \cdot \mathbf{A}'\}]_{\mathbf{K}}$ equals 0, or*

$$|\{[\{\mathbf{g} \cdot \mathbf{A}\}]_{\mathbf{P}} \cdot \mathbf{B}'\} - \{[\{\mathbf{g} \cdot \mathbf{B}\}]_{\mathbf{Q}} \cdot \mathbf{A}'\}| < \mathbf{K}^* .$$

Therefore,

$$[\{[\{\mathbf{g} \cdot \mathbf{A}\}]_{\mathbf{P}}\} \cdot \mathbf{B}']_{\mathbf{K}} = [\{[\{\mathbf{g} \cdot \mathbf{A}\}]_{\mathbf{Q}}\} \cdot \mathbf{B}']_{\mathbf{K}} + \Delta ,$$

where $\Delta = (\varepsilon_1/K_1, \varepsilon_2/K_2, \ldots, \varepsilon_n/K_n)$ and where each ε_i belongs to the set $\{-1, 0, 1\}$. In particular, the error vector Δ can take 3^n possible values.

For the proof of Theorem 2.2 see Appendix B.

Rounded GKE I Scheme

Setup (non-secret parameters):
- a natural number n
- natural n-tuples $\mathbf{P}, \mathbf{Q}$, and $\mathbf{K}$ as parameters of rounding
- mutually commuting subrings A and B of $M_n(\mathbb{Z})$

Protocol
- *Alice* and *Bob*: choose a non-secret $\mathbf{g} \in X_n$.

- *Alice*: choose a secret $\mathbf{a} = (a_1, a_2, \ldots, a_k) \in \mathbb{Z}^k$ such that $\mathbf{Q}^* \cdot |\varphi'_A(\mathbf{a})| \leq \mathbf{K}^*$
- *Bob*: choose a secret $\mathbf{b} = (b_1, b_2, \ldots, b_l) \in \mathbb{Z}^l$ such that $\mathbf{P}^* \cdot |\varphi'_B(\mathbf{b})| \leq \mathbf{K}^*$
- *Alice→Bob*: $m_A = [\{\mathbf{g} \cdot \varphi_A(\mathbf{a})\}]_\mathbf{P}$
- *Bob*: compute $S_B = [\{m_A \cdot \mathbf{b}\}]_\mathbf{K} = [\{[\{\mathbf{g} \cdot \varphi_A(\mathbf{a})\}]_\mathbf{P} \cdot \varphi'_B(\mathbf{b})\}]_\mathbf{K}$
- *Bob→ Alice*: $m_B = [\{\mathbf{g} \cdot \varphi_B(\mathbf{b})\}]_\mathbf{Q}$
- *Alice*: compute $S_A = [\{m_B \cdot \varphi'_A(\mathbf{a})\}]_\mathbf{K} = [\{[\{\mathbf{g} \cdot \varphi_B(\mathbf{b})\}]_\mathbf{P} \cdot \varphi'_A(\mathbf{a})\}]_\mathbf{K}$

Common Secret

By Theorem 2.2, we have $S_A = S_B + (\varepsilon_1/K_1, \varepsilon_2/K_2, \ldots, \varepsilon_n/K_n)$, where each $\varepsilon_i \in \{-1, 0, 1\}$. In particular, the difference between S_A and S_B can take at most 3^n values. This difference can be eliminated in the follow-up communication of Alice and Bob. Thus, the shared secret is the vector S_A.

REMARK 2.3. We express our gratitude to Gene Itkis for the idea to eliminate the difference between S_A and S_B using the follow-up communication of Alice and Bob.

Security of Rounded GKE I

Security of Rounded GKE I is based on hardness of the following is an analogue of the Discrete Logarithm Problem.

Given:

- a natural number n
- a subset $\mathbf{D}$ of $M_n(\mathbb{Z})$
- a natural n-tuple $\mathbf{P}$ and $\mathbf{g} \in X_n = [0, 1)^n$
- a vector $\mathbf{x} = [\{\mathbf{g} \cdot \mathbf{A}\}]_\mathbf{P}$ for some *unknown* $\mathbf{A} \in \mathbf{D}$

Compute: $\mathbf{A}' \in \mathbf{D}$ such that $\{\mathbf{g} \cdot \mathbf{A}'\}_\mathbf{P} = \mathbf{x}$

Clearly, the larger is the set $\mathbf{D}$ the harder is the problem. However, there is a natural limitation on the size of $\mathbf{D}$ because of the requirement of quasi-commutation of matrices $\mathbf{A}$ and $\mathbf{B}$. This and other GKE I – related problems will be analyzed in the work [4].

Numerical example

We take in the setup as above:

- $n = 2$
- $\mathbf{P} = (10^{18}, 10^{18})$, $\mathbf{K} = (10^{10}, 10^{10})$ as the parameters of rounding
- $\mathbf{M} = (10^8, 10^8)$.
- $A = B = \mathbb{Z}^2$ and for $\mathbf{a} = (a_0, a_1) \in \mathbb{Z}^2$ the 2×2 matrices $\varphi_A(\mathbf{a})$, $\varphi'_A(\mathbf{a})$, $\varphi_B(\mathbf{a})$, $\varphi'_B(\mathbf{a})$ are given by:

$$\varphi_A(\mathbf{a}) = \varphi'_A(\mathbf{a}) = \varphi_B(\mathbf{a}) = \varphi'_B(\mathbf{a}) = \begin{pmatrix} a_0 & -a_1 \\ a_1 & a_0 \end{pmatrix}.$$

As defined, these matrices satisfy the quasi-commutation equation (1.4).

Public *continuous* parameter: $\mathbf{g} = (g_1, g_2) = (\sqrt{2}, \sqrt{3})$.

Protocol

Alice chooses a pair of secret integers $(a_0, a_1) = (48176925, 18034725)$. Alice calculates the rounded vector

$$\mathbf{y} = (y_1, y_2) = ([\{g_1 a_0 + g_2 a_1\}]_\mathbf{P}, [\{-g_1 a_1 + g_2 a_0\}]_\mathbf{P}).$$

That is,
$$\mathbf{y} = ([\{\sqrt{2} \cdot 48176925 + \sqrt{3} \cdot 18034725\}]_{\mathbf{P}}, [\{-\sqrt{2} \cdot 18034725 + \sqrt{3} \cdot 48176925\}]_{\mathbf{P}}) =$$
$$= ([\{68132460.72843142218399029753959 6+31237060.000532620547511774721314\}]_{\mathbf{P}},$$
$$[\{-25504952.688669116604000035676723+83444881.852435233704474767836253\}]_{\mathbf{P}})$$
$$= (0.728964042731502072, 0.163766117100474732).$$

Each coordinate y_1, y_2 of this $\mathbf{y}$ has exactly 18 digits because $\mathbf{P} = (10^{18}, 10^{18})$. Alice sends this rounded vector $\mathbf{y}$ to Bob. Independently Bob chooses a pair of secret integers $(b_0, b_1) = (19082792, 27045821)$. Bob calculates the vector
$$\mathbf{z} = (z_1, z_2) = ([\{g_1 b_0 + g_2 b_1\}]_{\mathbf{P}}, [\{-g_1 b_1 + g_2 b_0\}]_{\mathbf{P}}).$$

That is,
$$\mathbf{z} = ([\{\sqrt{2} \cdot 19082792 + \sqrt{3} \cdot 27045821\}]_{\mathbf{P}}, [\{-\sqrt{2} \cdot 27045821 + \sqrt{3} \cdot 19082792\}]_{\mathbf{P}}) =$$
$$= ([\{26987143.254344799212512475172839+46844736.104413300451707772339473\}]_{\mathbf{P}},$$
$$[\{-38248566.863715063905876737732694+33052365.294268911065907204826118\}]_{\mathbf{P}}) =$$
$$= (0.358758099664220248, 0.430553847160030467)$$
and sends this vector $\mathbf{z}$ to Alice.

Upon receiving the vector $\mathbf{y}$ from Alice, Bob calculates the rounded vector $\mathbf{k} = (k_1, k_2)$ by the formula:
$$\mathbf{k} = (k_1, k_2) = ([\{y_1 b_0 + y_2 b_1\}]_{\mathbf{K}}, [\{-y_1 b_1 + y_2 b_0\}]_{\mathbf{K}}).$$

That is,
$$\mathbf{k} = ([\{0.728964042731502072 \cdot 19082792 + 0.163766117100474732 \cdot 27045821\}]_{\mathbf{K}},$$
$$[\{-0.728964042731502072 \cdot 27045821 + 0.163766117100474732 \cdot 19082792\}]_{\mathbf{K}}) =$$
$$= ([\{13910669.202924365887545024 + 4429189.088964478616694972\}]_{\mathbf{K}},$$
$$[\{-19715431.015152556100441112 + 3125114.749276002412011744\}]_{\mathbf{K}})$$
$$= (0.2918888445, 0.7341234463).$$

Each coordinate k_1, k_2 of this $\mathbf{k}$ has exactly 10 digits because $\mathbf{K} = (10^{10}, 10^{10})$.

Upon receiving the vector $\mathbf{z}$ from Bob, Alice calculates the rounded vector $\mathbf{k}' = (k'_1, k'_2)$ by the formula:
$$\mathbf{k}' = (k'_1, k'_2) = ([\{z_1 \cdot a_0 + z_2 \cdot a_1\}]_{\mathbf{K}}, [\{-z_1 \cdot a_1 + z_2 \cdot a_0\}]_{\mathbf{K}}).$$

That is,
$$\mathbf{k}' = ([\{0.358758099664220248 \cdot 48176925 + 0.430553847160030467 \cdot 18034725\}]_{\mathbf{K}},$$
$$[\{-0.358758099664220248 \cdot 18034725 + 0.430553847160030467 \cdot 48176925\}]_{\mathbf{K}}) =$$
$$= ([\{17283862.0606656640713774 + 7764920.231223180463966575\}]_{\mathbf{K}},$$
$$[\{-6470103.6689668045121118 + 20742760.403090250806373975\}]_{\mathbf{K}}) =$$
$$= (0.2918888445, 0.7341234463).$$

Thus, the vector $(0.2918888445, 0.7341234463)$ is the secret shared by Alice and Bob.

REMARK 2.4. Unlike in the general case of GKE, in this example Alice and Bob did not need any follow-up communication in order to establish the common secret out of $\mathbf{k}$ and $\mathbf{k}'$. They know that $\mathbf{k} = \mathbf{k}'$ because, on the one hand, Theorem 2.2 guarantees each coordinate of the difference $\mathbf{k} - \mathbf{k}'$ can be either 0 or $\pm 10^{-10}$ and, on the other hand, for each coordinate of each vector $\mathbf{k}$ and $\mathbf{k}'$ the 10th digit is neither 0 nor 9.

3. Geometric Key Establishment II

In this section we present a key establishment scheme based on Main Example II and on the general commuting double action key establishment scheme of Section 1. We will refer to it as geometric key establishment II (GKE II).

First, we present the ideal GKE II scheme (i.e., without any rounding).

Ideal Geometric Key Establishment II (GKE II) Scheme

Setup (public parameters):

- m and n are natural numbers
- $X_{m \times n} = M_{m \times n}([0,1))$ is the set of all $m \times n$ matrices with coefficients in the semi-open interval $[0,1)$. Each point of $X_{m \times n}$ is an $m \times n$ matrix $\mathbf{g} = (g_{ij})$, where each $g_{ij} \in [0,1)$

For each real $m \times n$ matrix $x = (x_{ij})$ we use the notation $\{x\} = (\{x_{ij}\})$, where $\{z\}$ stands for the fractional part of a real number z.

LEMMA 3.1. *For any integer matrices* $\mathbf{A} \in M_m(\mathbb{Z})$, $\mathbf{B} \in M_n(\mathbb{Z})$, *and any* $\mathbf{g} \in X_{m \times n}$ *one has*

(3.1) $$\{\{\mathbf{A} \cdot \mathbf{g}\} \cdot \mathbf{B}\} = \{\mathbf{A} \cdot \mathbf{g} \cdot \mathbf{B}\} = \{\mathbf{A} \cdot \{\mathbf{g} \cdot \mathbf{B}\}\} \, .$$

Proof. We will reduce the statement to Lemma 1.15. It suffices to show (similarly to the proof of Lemma 2.1) that the set $G = [0,1)$ is a group. Indeed, we have already shown that in the proof of Lemma 2.1. This proves the lemma. $\square$

Protocol

- *Alice* and *Bob*: choose a non-secret $\mathbf{g} \in X_{m \times n}$
- *Alice*: choose a secret $\mathbf{A} \in M_m(\mathbb{Z})$
- *Bob*: choose a secret $\mathbf{B} \in M_n(\mathbb{Z})$
- *Alice*→*Bob*: $m_A = \{\mathbf{A} \cdot \mathbf{g}\}$
- *Bob*: compute $S_B = \{m_A \cdot \mathbf{B}\} = \{\{\mathbf{A} \cdot \mathbf{g}\} \cdot \mathbf{B}\}$
- *Bob*→*Alice*: $m_B = \{\mathbf{g} \cdot \mathbf{B}\}$
- *Alice*: compute $S_A = \{\mathbf{A} \cdot m_B\} = \{\mathbf{A} \cdot \{\mathbf{g} \cdot \mathbf{B}\}\}$

Common Secret

By Lemma 3.1, $S_A = S_B$.

In order to present the *rounded* GKE II, we need the following notation. For any real $m \times n$ matrices $y = (y_{ij})$ and $z = (y_{ij})$, the matrix inequality $y \leq z$ is equivalent to $m \times n$ scalar inequalities: $y_{ij} \leq z_{ij}$ for $i = 1, 2, \ldots, m; j = 1, 2, \ldots, n$. Also the inequality $|y| < z$ means that $y < z$ and $-y < z$.

For a natural $m \times n$ matrix $\mathbf{P} = (P_{ij})$ and a real $m \times n$ matrix $\mathbf{g} = (g_{ij})$ such that each $g_{ij} \in [0,1)$, define the **P**-rounding to a rational $m \times n$ matrix $[\mathbf{g}]_{\mathbf{P}} \in X_{m \times n}$ by the formula:

$$[\mathbf{g}]_{\mathbf{P}} = ([g_{ij}]_{P_{ij}}) \, ,$$

where $[g]_P$ is defined in (2.2). For a natural $m \times n$ matrix $\mathbf{P} = (P_{ij})$ we denote $\mathbf{P}^* = (\frac{1}{P_{ij}})$.

THEOREM 3.2. *Let be $\mathbf{P}$, $\mathbf{Q}$, and $\mathbf{K}$ be natural $m \times n$ matrices. Then for any integer $m \times m$ matrix $\mathbf{A}$ and any integer $n \times n$ matrix $\mathbf{B}$ such that $|\mathbf{A}| \cdot \mathbf{Q}^* \leq \mathbf{K}^*$, $\mathbf{P}^* \cdot |\mathbf{B}| \leq \mathbf{K}^*$ one has: either at least one coefficient of the matrix $[\{[\{\mathbf{A} \cdot \mathbf{g}\}]_\mathbf{P} \cdot \mathbf{B}\}]_\mathbf{K}$ equals 0, or at least one coefficient of the matrix $[\{\mathbf{A} \cdot [\{\mathbf{g} \cdot \mathbf{B}\}]_\mathbf{Q}\}]_\mathbf{K}$ equals 0, or*

$$|\{[\{\mathbf{A} \cdot \mathbf{g}\}]_\mathbf{P} \cdot \mathbf{B}\} - \{\mathbf{A} \cdot [\{\mathbf{g} \cdot \mathbf{B}\}]_\mathbf{Q}\}| < \mathbf{K}^* .$$

Therefore,

$$[\{[\{\mathbf{A} \cdot \mathbf{g}\}]_\mathbf{P} \cdot \mathbf{B}\}]_\mathbf{K} = [\{\mathbf{A} \cdot [\{\mathbf{g} \cdot \mathbf{B}\}]_\mathbf{Q}\}]_\mathbf{K} + \Delta ,$$

where $\Delta = (\varepsilon_{ij}/K_{ij})$ and where each ε_{ij} belongs to the set $\{-1, 0, 1\}$. In particular, the error matrix Δ can take 3^{mn} possible values.

For the proof of Theorem 3.2 see Appendix B.

Rounded GKE II Scheme

Setup (public parameters):
- natural numbers m and n
- natural $m \times n$ matrices $\mathbf{P}$, $\mathbf{Q}$, and $\mathbf{K}$ as parameters of rounding

Protocol
- *Alice* and *Bob*: choose a non-secret $\mathbf{g} \in X_{m \times n}$
- *Alice*: choose a secret $\mathbf{A} \in M_m(\mathbb{Z})$ such that $|\mathbf{A}| \cdot \mathbf{Q}^* \leq \mathbf{K}^*$
- *Bob*: choose a secret $\mathbf{B} \in M_n(\mathbb{Z})$ such that $\mathbf{P}^* \cdot |\mathbf{B}| \leq \mathbf{K}^*$
- *Alice→Bob*: $m_A = [\{\mathbf{A} \cdot \mathbf{g}\}]_\mathbf{P}$
- *Bob*: compute $S_B = [\{m_A \cdot \mathbf{B}\}]_\mathbf{K} = [\{[\{\mathbf{A} \cdot \mathbf{g}\}]_\mathbf{P} \cdot \mathbf{B}\}]_\mathbf{K}$
- *Bob→Alice*: $m_B = [\{\mathbf{g} \cdot \mathbf{B}\}]_\mathbf{Q}$
- *Alice*: compute $S_A = [\{\mathbf{A} \cdot m_B\}]_\mathbf{K} = [\{\mathbf{A} \cdot \{\mathbf{g} \cdot \mathbf{B}\}]_\mathbf{P}\}]_\mathbf{K}$

Common Secret

By Theorem 3.2, one has $S_A = S_B + (\varepsilon_{ij}/K_{ij})$, where each $\varepsilon_{ij} \in \{-1, 0, 1\}$. In particular, the difference between S_A and S_B can take at most $3^{m \cdot n}$ values. This difference can be eliminated in the follow-up communication of Alice and Bob. Thus, the shared secret is the vector S_A.

The problem of security of Rounded GKE II follows the discussed above pattern of Rounded GKE I and will also be analyzed in [4].

Numerical example

We take in the setup as above:
- $m = n = 2$
- $\mathbf{P} = (P_{ij})$, where each $P_{ij} = 10^9$, $\mathbf{K} = (K_{ij})$ where each $K_{ij} = 10^5$
- $\mathbf{M} = (M_{ij})$, where each $M_{ij} = 10^3$

Public *continuous* parameter:

$$\mathbf{g} = (g_{ij}) = \begin{pmatrix} \sqrt{2} & \sqrt{3} \\ \sqrt{5} & \sqrt{7} \end{pmatrix}$$

Protocol

Suppose that Alice chooses a secret integer 2×2 matrix $\mathbf{A}$:

$$\mathbf{A} = \begin{pmatrix} 123 & 456 \\ 817 & 391 \end{pmatrix} .$$

Alice calculates the 2×2 matrix $\mathbf{y} = [\{\mathbf{A} \cdot \mathbf{g}\}]$ each element of which rounded to 9 decimal places:

$$\mathbf{y} = [\{\mathbf{A} \cdot \mathbf{g}\}] = \begin{pmatrix} 0.595265912 & 0.504847176 \\ 0.715059661 & 0.574272410 \end{pmatrix}$$

and sends this 2×2 matrix $\mathbf{y}$ to Bob. Suppose that at independently Bob chooses a secret integer 2×2 matrix $\mathbf{B}$:

$$\mathbf{B} = \begin{pmatrix} 691 & 378 \\ 529 & 109 \end{pmatrix}.$$

Bob calculates the 2×2 matrix $\mathbf{z} = [\{\mathbf{g} \cdot \mathbf{B}\}]$ each element of which rounded to 9 decimal places:

$$\mathbf{z} = [\{\mathbf{g} \cdot \mathbf{B}\}] = \begin{pmatrix} 0.476448804 & 0.366264602 \\ 0.725416006 & 0.620588401 \end{pmatrix}$$

and sends this 2×2 matrix $\mathbf{z}$ to Alice. Upon receiving the 2×2 matrix $\mathbf{y}$ from Alice, Bob calculates the 2×2 matrix $\mathbf{k} = [\{\mathbf{y} \cdot \mathbf{B}\}]$ with the precision 5 decimal places after dot:

$$\mathbf{k} = [\{\mathbf{y} \cdot \mathbf{B}\}] = \begin{pmatrix} 0.39290 & 0.03885 \\ 0.89633 & 0.88824 \end{pmatrix}.$$

Upon receiving the 2×2 matrix $\mathbf{z}$ from Bob, Alice calculates the 2×2 matrix $\mathbf{k}' = [\{\mathbf{A} \cdot \mathbf{z}\}]$ with the precision 5 decimal places after dot:

$$\mathbf{k}' = [\{\mathbf{A} \cdot \mathbf{z}\}] = \begin{pmatrix} 0.39290 & 0.03885 \\ 0.89633 & 0.88824 \end{pmatrix}.$$

REMARK 3.3. Unlike in the general case of GKE II, in this example Alice and Bob did not need any follow-up communication in order to establish the common secret out of $\mathbf{k}$ and $\mathbf{k}'$. They know that $\mathbf{k} = \mathbf{k}'$ because, on the one hand, Theorem 3.2 guarantees that each matrix coefficient of the difference $\mathbf{k} - \mathbf{k}'$ can be either 0 or $\pm 10^{-5}$ and, on the other hand, for each coefficient of each matrix $\mathbf{k}$ and $\mathbf{k}'$ the 5th digit is neither 0 nor 9.

4. Appendix A. General Key Establishment Scheme and its rounded versions

We start with a natural generalization of Diffie-Hellman protocol. Apparently, this generalization is well known, but we have failed to find references. Hence, we will take liberty to call it 'basic key establishment scheme.'

DEFINITION 4.1. Let A, B and X, Y_A, Y_B, Z be sets. Let $A \times X \to Y_A$, $B \times Y_A \to Z$, and $B \times X \to Y_B$, $A \times Y_B \to Z$ be a quadruple of maps (we denote them respectively by $(\mathbf{a}, \mathbf{x}) \to \mathbf{a}(\mathbf{x})$, $(\mathbf{b}, \mathbf{y}) \to \mathbf{b}(\mathbf{y})$, and $(\mathbf{b}, \mathbf{x}) \to \mathbf{b}(\mathbf{x})$, $(\mathbf{a}, \mathbf{y}') \to \mathbf{a}(\mathbf{y}')$ for any elements $\mathbf{a} \in A$, $\mathbf{b} \in B$, $\mathbf{x} \in X$, $\mathbf{y} \in Y_A$, $\mathbf{y}' \in Y_B$). We say that the quadruple is *commuting* if:

(4.1) $$\mathbf{a}(\mathbf{b}(\mathbf{x})) = \mathbf{b}(\mathbf{a}(\mathbf{x}))$$

for any $\mathbf{a} \in A$, $\mathbf{b} \in B$, $\mathbf{x} \in X$.

The basic key establishment scheme consists of the following setup and protocol.

Basic key establishment scheme
Setup

- sets A and B (of private parameters)
- a set X (of shared parameters)
- a set Y_A (of Alice's transmittable elements)
- a set Y_B (of Bob's transmittable elements)
- a set Z (of shared secret elements)
- a commuting quadruple $A \times X \to Y_A$, $B \times Y_A \to Z$, $B \times X \to Y_B$, $A \times Y_B \to Z$

Protocol
- *Alice* and *Bob*: choose a non-secret $\mathbf{x} \in X$
- *Alice*: choose a secret element $\mathbf{a} \in A$
- *Bob*: choose a secret element $\mathbf{b} \in B$
- *Alice$\to$Bob*: $m_A = \mathbf{a}(\mathbf{x})$
- *Bob*: compute $S_B = \mathbf{b}(m_A) = \mathbf{b}(\mathbf{a}(\mathbf{x}))$
- *Bob$\to$Alice*: $m_B = \mathbf{b}(\mathbf{x})$
- *Alice*: compute $S_A = \mathbf{a}(m_B) = \mathbf{a}(\mathbf{b}(\mathbf{x}))$

Common Secret

By Definition 4.1, $S_A = S_B$.

REMARK 4.2. The scheme is secure if the following problem is hard: given $\mathbf{x} \in X$ and $\mathbf{y} \in Y_A$, find $\mathbf{a} \in A$ such that $\mathbf{y} = \mathbf{a}(\mathbf{x})$. In the case of the original Diffie-Hellman scheme ([5]), this problem is known as the *discrete logarithm problem*.

Of course, for the purpose of implementation of this basic scheme, it is natural to require that all the involved sets A, B, X, Y_A, Y_B, and Z are finite. In Sections 2 and 3 we presented a method for generation of a large family of finite key establishment schemes each of which represents a non-trivial approximation of the basic scheme. The richness of the family stems from its origin in an *infinite* or even *continuous* instantiation of the basic scheme.

Generalizing rounded schemes introduced in Sections 2 and 3, we propose here a general *rounded key establishment* (RKE) scheme. In the following definitions and results we need a mathematical concept of 'metric space'. For the standard references, see e.g., [6].

DEFINITION 4.3. A *metric space* is a pair (X, d), where X is a set and $d: X \times X \to R_{\geq 0}$ is a *distance function* on X satisfying:
- (symmetry) $d(\mathbf{x}, \mathbf{x}') = d(\mathbf{x}', \mathbf{x})$ for all $\mathbf{x}, \mathbf{x}' \in X$
- $d(\mathbf{x}, \mathbf{x}') = 0$ if and only if $\mathbf{x} = x'$
- (triangle inequality) $d(\mathbf{x}, \mathbf{x}'') \geq d(\mathbf{x}, \mathbf{x}') + d(x', \mathbf{x}'')$ for all $\mathbf{x}, \mathbf{x}', \mathbf{x}'' \in X$

DEFINITION 4.4. Let (X, d) and (Y, d) be metric spaces. Then a map $F: X \to Y$ is called *metric* if there exists a positive constant C such that $d(F(\mathbf{x}), F(x')) \leq C \cdot d(\mathbf{x}, \mathbf{x}')$ for any $\mathbf{x}, \mathbf{x}' \in X$. More generally, given a function $f: A \to R_{>0}$, we say that a map $A \times X \to Y$ (which we denote $(\mathbf{a}, \mathbf{x}) \to \mathbf{a}(\mathbf{x})$) is f-*Lipschitz* if $d(\mathbf{a}(\mathbf{x}), \mathbf{a}(x')) \leq f(\mathbf{a}) \cdot d(\mathbf{x}, \mathbf{x}')$ for any $\mathbf{x}, \mathbf{x}' \in X, \mathbf{a} \in A$.

DEFINITION 4.5. Let (X, d) be a metric space. Let K be a discrete subset of X. Consider a map $[\cdot]: X \to K$ (to be denoted by $\mathbf{x} \to [\mathbf{x}]$) such that for each $\mathbf{x} \in X$ the distance from $\mathbf{x} to [\mathbf{x}]$ does not exceed the distance from $\mathbf{x}$ to any other point of K. We refer to any such map as K-*rounding* on (X, d).

In what follows we will consider only infinite or even uncountable metric spaces.

DEFINITION 4.6. Let (X, d) be a metric space. Let us consider an infinite ascending chain
$$X_1 \subset X_2 \subset X_3 \subset \ldots \subset X_k \subset \ldots$$
of discrete subsets (each inclusion is strict), and let $\mathbf{r} = \{r_k\}$, $k = 1, 2, \ldots$ be a decreasing sequence of positive real numbers converging to 0. Given an infinite family $[\cdot]_k : X \to X_k$ of X_k-roundings on (X, d) for $k = 1, 2, \ldots$, we say that the family $[\cdot]_k$ is $\mathbf{r}$-*saturated* if $d(\mathbf{x}, [\mathbf{x}]_k) \leq r_k$ for any point $\mathbf{x}$ and for each natural number k.

DEFINITION 4.7. Let (X, d) be a metric space, $\mathbf{x}$ be a point of X, and r be a positive real number. Denote by $B(\mathbf{x}; r)$ the set of all points $\mathbf{x}' \in X$ such that $d(\mathbf{x}, \mathbf{x}') < r$. We refer to $B(\mathbf{x}; r)$ as the *open ball* of radius r centered at $\mathbf{x}$.

DEFINITION 4.8. Let (X, d) be a metric space, $\mathbf{r} = \{r_k\}$ be a sequence of positive real numbers, and N be a natural number. We say that an ascending chain
$$X_1 \subset X_2 \subset X_3 \subset \ldots \subset X_k \subset \ldots$$
of subsets of X is $(\mathbf{r}, N)$-*uniform* if $|B(\mathbf{x}; r_k) \cap X_k| < N$ for every $\mathbf{x} \in X_k$ and each $k = 1, 2, \ldots$.

Informally speaking, $(\mathbf{r}, N)$-uniform chains in X provide good approximations of points of X similarly to the way in which rational numbers provide good approximations of real numbers.

DEFINITION 4.9. For a given $(\mathbf{r}, N)$-uniform ascending chain
$$X_1 \subset X_2 \subset X_3 \subset \ldots \subset X_k \subset \ldots$$
in a metric space (X, d) we say that two points $\mathbf{k}$ and $\mathbf{k}'$ of X_k are *neighbors* if $d(\mathbf{k}, \mathbf{k}') < r_k$.

By definition, any point $\mathbf{k} \in X_k$ has at most N neighbors.

THEOREM 4.10. Let A, B, and X be sets, and $A \times X \to Y_A$, $B \times X \to Y_B$ be maps. Let (Y_A, d), (Y_B, d), and (Z, d) be metric spaces, and let $B \times Y_A \to Z$ be a g-Lipschitz map, $A \times Y_B \to Z$ be a g'-Lipschitz map. Also let $[\cdot]_m : Y_A \to (Y_A)_m$ be an $\mathbf{r}$-saturated rounding on (Y_A, d), $[\cdot]'_m : Y_B \to (Y_B)_m$ be an $\mathbf{r}'$-saturated family of roundings on (Y_B, d), and $[\cdot]''_k : Z \to (Z)_k$ be an $\mathbf{r}''$-saturated family of roundings on (Y_B, d) such that the ascending chain $Z_1 \subset Z_2 \subset Z_3 \subset \ldots \subset Z_k \subset \ldots$ is $(3\mathbf{r}'', N)$-uniform. Then for any commuting elements $\mathbf{a} \in A$ and $\mathbf{b} \in B$ such that $g(\mathbf{b}) < r''_k/(2r_m)$, $g'(\mathbf{a}) < r''_k/(2r'_m)$ (for some natural m, k) and any $\mathbf{x} \in X$ one has: $[\mathbf{a}([\mathbf{b}(\mathbf{x})]'_m)]''_k$ and $[\mathbf{b}([\mathbf{a}(\mathbf{x})]_m)]''_k$ are neighbors.

Proof. By definition of saturated families of roundings, one has for any m:
$$d(\mathbf{a}(\mathbf{x}), [\mathbf{a}(\mathbf{x})]_m) \leq r_m, d(\mathbf{b}(\mathbf{x}), [\mathbf{b}(\mathbf{x})]'_m) \leq r'_m .$$
Therefore, for given m and k we have:
$$d(\mathbf{b}(\mathbf{a}(\mathbf{x})), \mathbf{b}([\mathbf{a}(\mathbf{x})]_m)) \leq g(\mathbf{b}) \cdot d(\mathbf{a}(\mathbf{x}), [\mathbf{a}(\mathbf{x})]_m) \leq g(\mathbf{b}) \cdot r_m < r''_k/2 ,$$
$$d(\mathbf{a}(\mathbf{b}(\mathbf{x})), \mathbf{a}([\mathbf{b}(\mathbf{x})]'_m)) \leq g'(\mathbf{a}) \cdot d(\mathbf{b}(\mathbf{x}), [\mathbf{a}(\mathbf{x})]'_m) \leq g'(\mathbf{a}) \cdot r'_m < r''_k/2 .$$
Denote $\mathbf{z} = (\mathbf{a}(\mathbf{b}(\mathbf{x})) = (\mathbf{b}(\mathbf{a}(\mathbf{x}))$. Then
$$d(\mathbf{z}, \mathbf{b}([\mathbf{a}(\mathbf{x})]_m)) \leq r''_k/2, d(\mathbf{z}, \mathbf{a}([\mathbf{b}(\mathbf{x})]_m)) \leq r''_k/2 .$$

Denote $\mathbf{k}_1 = [\mathbf{a}([\mathbf{b}(\mathbf{x})]'_m)]''_k$ and $\mathbf{k}_2 = [\mathbf{b}([\mathbf{a}(\mathbf{x})]_m)]''_k$. Note that
$$d(\mathbf{k}_1, \mathbf{a}([\mathbf{b}(\mathbf{x})]'_m)) \leq r''_k, \ d(\mathbf{k}_2, \mathbf{b}([\mathbf{a}(\mathbf{x})]_m)) \leq r''_k \ .$$

Then, by the triangle inequality,
$$d(\mathbf{z}, \mathbf{k}_1) \leq d(\mathbf{z}, \mathbf{a}([\mathbf{b}(\mathbf{x})]_m)) + d(\mathbf{k}_1, \mathbf{a}([\mathbf{b}(\mathbf{x})]_m)) < r''_k/2 + r''_k = 3r''_k/2,$$
$$d(\mathbf{z}, \mathbf{k}_2) \leq d(\mathbf{z}, \mathbf{a}([\mathbf{b}(\mathbf{x})]_m)) + d(\mathbf{k}_2, \mathbf{b}([\mathbf{a}(\mathbf{x})]_m)) < r''_k/2 + r_k = 3r''_k/2 \ .$$

Finally, again by the triangle inequality,
$$d(\mathbf{k}_1, \mathbf{k}_2) \leq d(\mathbf{z}, \mathbf{k}_1) + d(\mathbf{z}, \mathbf{k}_2) < 3r''_k/2 + 3r''_k/2 = 3r''_k \ .$$

That is, $\mathbf{k_1}$ and $\mathbf{k_2}$ are neighbors. Theorem 4.10 is proved. $\square$

Based on this result we propose the following general *rounded* key establishment scheme.

Rounded key establishment (RKE) scheme
Setup
- sets A and B of private parameters
- a set X of shared parameters
- infinite metric spaces (Y_A, d) and (Y_B, d) of transmittable elements
- an infinite metric space (Z, d) of shared secret elements
- maps $A \times X \to Y_A$, $B \times X \to Y_B$, a g-Lipschitz map $B \times Y_A \to Z$, and a g'-Lipschitz map $A \times Y_B \to Z$ such that the quadruple of these maps is commuting
- an **r**-saturated family of roundings $[\cdot]_m : Y_A \to (Y_A)_m$ on (Y_A, d) and an **r'**-saturated family of roundings $[\cdot]'_m : Y_B \to (Y_B)_m$ on (Y_B, d)
- an **r''**-saturated family of roundings $[\cdot]''_k : Z \to (Z)_k$ on (Z, d) such that the ascending chain $Z_1 \subset Z_2 \subset Z_3 \subset \ldots \subset Z_k \subset \ldots$ is $(3\mathbf{r''}, N)$-*uniform*

Protocol
- *Alice* and *Bob*: choose a shared parameter $\mathbf{x} \in X$ and natural numbers m, k
- *Alice*: choose a private element $\mathbf{a} \in A$ such that $g'(\mathbf{a}) < r''_k/(2r'_m)$
- *Bob*: choose a private element $\mathbf{b} \in B$ such that $g(\mathbf{b}) < r''_k/(2r_m)$
- *Alice*→*Bob*: $m_A = [\mathbf{a}(\mathbf{x})]_m$
- *Bob*: compute $S_B = [\mathbf{b}(m_A)]''_k = [\mathbf{b}([\mathbf{a}(\mathbf{x})]_m)]''_k$
- *Bob*→*Alice*: $m_B = [\mathbf{b}(\mathbf{x})]'_m$
- *Alice*: compute $S_A = [\mathbf{a}(m_B)]''_k = [\mathbf{a}([\mathbf{b}(\mathbf{x})]'_m)]''_k$

Common Secret
By Theorem 4.10, S_A and S_B are neighbors. Therefore, after making at most N choices, and without revealing the elements they computed, Alice and Bob select S_A as their shared secret.

5. Appendix B. Proof of results of Sections 2 and 3

Proof of Theorem 2.2. By definition, one has:
$$[\{\mathbf{g} \cdot \mathbf{A}\}]_\mathbf{P} = \{\mathbf{g} \cdot \mathbf{A}\} + \theta_1, \ \{\mathbf{g} \cdot \mathbf{B}\}_\mathbf{Q} = \{\mathbf{g} \cdot \mathbf{B}\} + \theta_2,$$
where $-\frac{1}{2} \cdot \mathbf{P}^* \leq \theta_1 \leq \frac{1}{2} \cdot \mathbf{P}^*$ and $-\frac{1}{2} \cdot \mathbf{Q}^* \leq \theta_2 \leq \frac{1}{2} \cdot \mathbf{Q}^*$. Therefore,
$$[\{\mathbf{g} \cdot \mathbf{A}\}]_\mathbf{P} \cdot \mathbf{B}' = (\{\mathbf{g} \cdot \mathbf{A}\} + \theta_1) \cdot \mathbf{B}' = \{\mathbf{g} \cdot \mathbf{A}\} \cdot \mathbf{B}' + \theta_1 \cdot \mathbf{B}' = \{\mathbf{g} \cdot \mathbf{A}\} \cdot \mathbf{B}' + \mathbf{E}_1 \ ,$$
where $\mathbf{E}_1 = \theta_1 \cdot \mathbf{B}'$.

Similarly,

$$[\{\mathbf{g}\cdot\mathbf{B}\}]_\mathbf{Q}\cdot\mathbf{A}' = (\{\mathbf{g}\cdot\mathbf{B}\}+\theta_2)\cdot\mathbf{A}' = \{\mathbf{g}\cdot\mathbf{B}\}\cdot\mathbf{A}' + \theta_2\cdot\mathbf{A}' = \{\mathbf{g}\cdot\mathbf{B}\}\cdot\mathbf{A}' + \mathbf{E}_2,$$

where $\mathbf{E}_2 = \theta_2\cdot\mathbf{A}'$.

By the assumptions, one has:

$$|\mathbf{E}_1| = |\theta_1\cdot\mathbf{B}'| \leq \frac{1}{2}\cdot|P^*\cdot\mathbf{B}'| < \frac{1}{2}\cdot P^*\cdot\mathbf{B}' \leq \frac{1}{2}\cdot K^*, |\mathbf{E}_2| = |\theta_2\cdot\mathbf{A}'| \leq \frac{1}{2}\cdot|Q^*\cdot\mathbf{A}'| < \frac{1}{2}\cdot Q^*\cdot\mathbf{B}' \leq \frac{1}{2}\cdot K^*.$$

In its turn, the inequality $|\mathbf{E}_1| \leq \frac{1}{2}\cdot K^*$ implies that either the vector $\{\{\mathbf{g}\cdot\mathbf{A}\}_\mathbf{P}\cdot\mathbf{B}'\}_\mathbf{K}$ has a coordinate equal to 0 or:

$$\{[\{\mathbf{g}\cdot\mathbf{A}\}]_\mathbf{P}\cdot\mathbf{B}'\} = \{\{\mathbf{g}\cdot\mathbf{A}\}\cdot\mathbf{B}' + \mathbf{E}_1\} = \{\{\mathbf{g}\cdot\mathbf{A}\}\cdot\mathbf{B}'\} + \mathbf{E}_1 = \{\mathbf{g}\cdot\mathbf{A}\cdot\mathbf{B}'\} + \mathbf{E}_1\ .$$

Similarly, the inequality $|\mathbf{E}_2| \leq \frac{1}{2}\cdot K^*$ implies that either the vector $\{\{\mathbf{g}\cdot\mathbf{B}\}_\mathbf{Q}\cdot\mathbf{A}'\}_\mathbf{K}$ has a coordinate equal to 0 or:

$$\{[\{\mathbf{g}\cdot\mathbf{B}\}]_\mathbf{Q}\cdot\mathbf{A}'\} = \{\{\mathbf{g}\cdot\mathbf{B}\}\cdot\mathbf{A}' + \mathbf{E}_2\} = \{\{\mathbf{g}\cdot\mathbf{B}\}\cdot\mathbf{A}'\} + \mathbf{E}_2 = \{\mathbf{g}\cdot\mathbf{B}\cdot\mathbf{A}'\} + \mathbf{E}_2\ .$$

Since $\mathbf{A}\cdot\mathbf{B}' = \mathbf{B}\cdot\mathbf{A}'$, one has $\{[\{\mathbf{g}\cdot\mathbf{A}\}]_\mathbf{P}\cdot\mathbf{B}'\} - \{[\{\mathbf{g}\cdot\mathbf{B}\}]_\mathbf{Q}\cdot\mathbf{A}'\} = \mathbf{E}_1 - \mathbf{E}_2$. Finally note that

$$|\mathbf{E}_1 - \mathbf{E}_2| \leq |\mathbf{E}_1| + |\mathbf{E}_2| < \frac{1}{2}\cdot\mathbf{K}^* + \frac{1}{2}\cdot\mathbf{K}^* = \mathbf{K}^*\ .$$

Theorem 2.2 is proved. $\square$

Proof of Theorem 3.2. By definition, one has

$$[\{\mathbf{A}\cdot\mathbf{g}\}]_\mathbf{P} = \{\mathbf{A}\cdot\mathbf{g}\} + \theta_1,\ [\{\mathbf{g}\cdot\mathbf{B}\}]_\mathbf{Q} = \{\mathbf{g}\cdot\mathbf{B}\} + \theta_2\ ,$$

where θ_1 and θ_2 are real $m\times n$ matrices such that

$$-\frac{1}{2}\mathbf{P}^* \leq \theta_1 \leq \frac{1}{2}\mathbf{P}^*\ \text{and}\ -\frac{1}{2}\mathbf{Q}^* \leq \theta_2 \leq \frac{1}{2}\mathbf{Q}^*\ .$$

Therefore,

$$([\{\mathbf{A}\cdot\mathbf{g}\}]_\mathbf{P})\cdot\mathbf{B} = (\{\mathbf{A}\cdot\mathbf{g}\}+\theta_1)\cdot\mathbf{B} = \{\mathbf{A}\cdot\mathbf{g}\}\cdot B + \theta_1\cdot\mathbf{B} = \{\mathbf{A}\cdot\mathbf{g}\}\cdot\mathbf{B} + \mathbf{E}_1, where \mathbf{E}_1 = \theta_1\cdot\mathbf{B}\ .$$

Similarly,

$$\mathbf{A}\cdot([\{\mathbf{g}\cdot\mathbf{B}\}]_\mathbf{Q}) = \mathbf{A}\cdot(\{\mathbf{g}\cdot\mathbf{B}\}+\theta_2\cdot\mathbf{Q}^*) = \mathbf{A}\cdot\{\mathbf{g}\cdot\mathbf{B}\} + \mathbf{A}\cdot\theta_2 = \mathbf{A}\cdot\{\mathbf{g}\cdot\mathbf{B}\} + \mathbf{E}_2\ ,$$

where $\mathbf{E}_2 = \mathbf{A}\cdot\theta_2$.

By the assumptions, one has:

$$|\mathbf{E}_1| = |\theta_1\cdot\mathbf{B}| \leq \frac{1}{2}\cdot|\mathbf{P}^*\cdot\mathbf{B}| < \frac{1}{2}\cdot\mathbf{K}^*,\ |\mathbf{E}_2| = |\mathbf{A}\cdot\theta_2| \leq \frac{1}{2}\cdot|\mathbf{A}\cdot\mathbf{Q}^*| < \frac{1}{2}\cdot\mathbf{K}^*\ .$$

In its turn, this implies that either at least one coefficient of the matrix $[\{[\{\mathbf{A}\cdot\mathbf{g}\}]_\mathbf{P}\cdot\mathbf{B}\}]_\mathbf{K}$ equals 0, or at least one coefficient of the matrix $[\{\mathbf{A}\cdot[\{\mathbf{g}\cdot\mathbf{B}\}]_\mathbf{Q}\}]_\mathbf{K}$ equals 0, or:

$$\{([\{\mathbf{A}\cdot\mathbf{g}\}]_\mathbf{P})\cdot\mathbf{B}\} = \{\{\mathbf{A}\cdot\mathbf{g}\}\cdot\mathbf{B} + \mathbf{E}_1\} = \{\{\mathbf{A}\cdot\mathbf{g}\}\cdot\mathbf{B}\} + \mathbf{E}_1 = \{\mathbf{A}\cdot\mathbf{g}\cdot\mathbf{B}\} + \mathbf{E}_1\ .$$

Similarly, the above implies that either at least one coefficient of the matrix $[\{\mathbf{A}\cdot[\{\mathbf{g}\cdot\mathbf{B}\}]_\mathbf{Q}\}]_\mathbf{K}$ is 0 or:

$$\{\mathbf{A}\cdot([\{\mathbf{g}\cdot\mathbf{B}\}]_\mathbf{Q})\} = \{\mathbf{A}\cdot\{\mathbf{g}\cdot\mathbf{B}\} + \mathbf{E}_2\} = \{\mathbf{A}\cdot\{\mathbf{g}\cdot\mathbf{B}\}\} + \mathbf{E}_2 = \{\mathbf{A}\cdot\mathbf{g}\cdot\mathbf{B}\} + \mathbf{E}_2\ .$$

Therefore

$$\{([\{\mathbf{A}\cdot\mathbf{g}\}]_\mathbf{P})\cdot\mathbf{B}\} - \{\mathbf{A}\cdot([\{\mathbf{g}\cdot\mathbf{B}\}]_\mathbf{Q})\} = \mathbf{E}_1 - \mathbf{E}_2\ .$$

Finally note that
$$|\mathbf{E}_1 - \mathbf{E}_2| \leq |\mathbf{E}_1| + |\mathbf{E}_2| < \frac{1}{2} \cdot \mathbf{K}^* + \frac{1}{2} \cdot \mathbf{K}^* = \mathbf{K}^* \ .$$

Theorem 3.2 is proved.

References

[1] I. Anshel, M. Anshel, and D. Goldfeld, An algebraic method for public-key cryptography, *Mathematical Research Letters* **6**, (1999). 1-5.

[2] I. Anshel, M. Anshel, and D. Goldfeld, A Linear Time Matrix Key Agreement Protocol, preprint, *http://www.ipam.ucla.edu/publications/cry2002/cry2002_dgoldfeld.pdf*

[3] I. Anshel, M. Anshel, and D. Goldfeld, "Non-abelian key agreement protocols," Discrete Appl. Math. 130 (2003), no. 1, 3-12.

[4] A. Berenstein, L. Chernyak, G. Itkis, ArKE: "Arithmetic Key Establishment System and its Security," in preparation.

[5] W. Diffie and M. E. Hellman, "New Directions in Cryptography," IEEE Transaction on Information Theory vol. IT 22 (November 1976), pp. 644-654.

[6] A. Kolmogorov, and S. Fomin, *Introductory real analysis*. Dover Publications, Inc., New York, 1975.

[7] G. Maze, C. Monico, and J. Rosenthal. "A public key cryptosystem based on actions by semigroups." In Proceedings of the 2002 IEEE International Symposium on Information Theory, page 266, Lausanne, Switzerland, 2002.

[8] V. Sidel'nikov, M. Cherepnev, V. Yashchenko, Systems of open distribution of keys on the basis of noncommutative semigroups, Russian Acad. Sci. Dokl. Math., vol. 48 (1994), No. 2, 384-386.

DEPARTMENT OF MATHEMATICS, UNIVERSITY OF OREGON, EUGENE, OR 97403, USA
E-mail address: `arkadiy@math.uoregon.edu`

INSTITUTE FOR SELF-ORGANIZING SYSTEMS, LLC. BOSTON, MASSACHUSETTS
E-mail address: `lchernyak@comcast.net`

Using shifted conjugacy in braid-based cryptography

Patrick DEHORNOY

ABSTRACT. Conjugacy is not the only possible primitive for designing braid-based protocols. To illustrate this principle, we describe a Fiat–Shamir-style authentication protocol that be can be implemented using any binary operation that satisfies the left self-distributive law. Conjugation is an example of such an operation, but there are other examples, in particular the shifted conjugation on Artin's braid group B_∞, and the finite Laver tables. In both cases, the underlying structures have a high combinatorial complexity, and they lead to difficult problems.

Most of the braid-based cryptographic schemes proposed so far [1, 18, 3] rely on the supposed complexity of the conjugation operation in Artin's braid groups. In this note, we would like to stress the fact that conjugation is by far not the only possible primitive operation for designing braid-based protocols.

To illustrate this general idea on a concrete example, we shall discuss an authentication scheme directly reminiscent of the Fiat–Shamir scheme, and a variant of some scheme considered in [20] in the case of braids. We show that such a scheme can be implemented naturally in every algebraic system that involves a binary operation that satisfies the algebraic law $x(yz) = (xy)(xz)$, called (left) self-distributivity. Conjugation on any group is an example of such an operation, but there are other examples, in particular the operation that we call *shifted conjugation* on Artin's braid group B_∞. There are reasons to think that sfifted conjugation is (much) more complicated than standard conjugation, and it could provide a promising alternative primitive for braid-based cryptography.

We also mention the Laver tables, which provide other examples of self-distributive operations, this time on a finite underlying domain of size 2^n. Again, these combinatorially very complex structures could provide a valuable platform.

1. A Fiat–Shamir-like authentication scheme

Here we start with the general principle of the Fiat–Shamir authentication scheme, and show that, under rather natural hypotheses, it can be implemented in any algebraic system involving a self-distributive binary operation.

1.1. The general principle. Let us start with an arbitrary set S, and try to construct an authentication scheme using the elements of S. To this end, we assume that a function F_s of S into itself is attached to each element of S and that there exist efficiently sampleable distributions on S such that, provided s and p are chosen according to them, the probability that s can be retrieved from the pair

---

1991 *Mathematics Subject Classification.* 94A60, 94A62, 68P25, 20F36.

Key words and phrases. cryptography; self-distributive operation; braid group; Fiat–Shamir scheme; Laver table; central duplication.

$(p, F_s(p))$ in feasible running time is negligible. Under such hypotheses, we can use s as a private key, and $(p, F_s(p))$ as a public key.

A natural idea for designing an authentication scheme is to let the prover appeal to a second, auxiliary (random) key r, and use $F_r(s)$ as a disguised version of s. What we need for a Fiat–Shamir-like authentication scheme is a commitment of the verifier guaranteeing that r is fixed, *and* an equality witnessing that $F_r(s)$ is connected in some way to s, *via* the commitment of the prover. As the elements p and $F_s(p)$ are public, it is natural to use $F_r(p)$ and/or $F_r(F_s(p))$ as the commitment(s) of the prover. Indeed, the assumption that x cannot be retrieved from $(y, F_x(y))$, which is already needed for $(p, F_s(p))$, automatically guarantees that r cannot be retrieved from the commitments of the prover.

Then what we need is some equality connecting $x = F_r(p)$, $y = F_r(F_s(p))$, and s—in a way that heavily involves s, *i.e.*, in such a way that the probability for another $\tilde{s}$ to give rise to the same equality is negligible. A simple, but very particular, solution is to require that F_s and F_r commute: in this case, the connection between x and y is just $y = F_s(x)$. This situation is essentially that considered in [**21, 18**], and it is not suitable in the current framework as the verifier would have to know the secret s.

A more general and flexible solution is to require that $F_r(F_s(p))$ be connected to $F_r(p)$ and s by some relation of the form $F_r(F_s(p)) = G_{r,s}(F_r(p))$ for some new function $G_{r,s}$. A not so special case is when $G_{r,s}$ is itself of the form $F_{g(r,s)}$ where g is some mapping of $S \times S$ into S: considering such a case is natural, because it avoids introducing a new family of functions and it enables one to work with the functions $(F_s)_{s \in S}$ solely. For the same reason, it is natural to consider the case when $g(r, s)$ is defined in terms of the F-functions, typically $g(r, s) = F_r(s)$. This leads to requiring that the functions F_s satisfy the condition

$$(1.1) \qquad F_r(F_s(p)) = F_{F_r(s)}(F_r(p)),$$

and to use this equality for proving authentication.

1.2. An authentication scheme. The previous analysis leads to considering the following authentication scheme.

We assume that S is a set and $(F_s)_{s \in S}$ is a family of functions of S to itself that satisfies Condition (1.1). Then the public keys are a pair (p, p') of elements of S satisfying $p' = F_s(p)$, while s is Alice's private key. The authentication procedure consists in repeating k times the following three exchanges:

A chooses r in S, and sends the *commitments* $x = F_r(p)$ and $x' = F_r(p')$;

B chooses a random bit c and sends it to A;

For $c = 0$, A sends $y = r$, and B checks $x = F_y(p)$ and $x' = F_y(p')$;

For $c = 1$, A sends $y = F_r(s)$, and B checks $x' = F_y(x)$.

The correctness of the scheme directly follows from Condition (1.1). Its security relies on the following assumptions:

(*i*) It is impossible to retrieve s from the pair $(p, F_s(p))$, and, more generally, it is impossible to find $\tilde{s}$ satisfying $F_s(p) = F_{\tilde{s}}(p)$; similar assertions hold for the pairs $(p, F_r(p))$, $(p', F_r(p'))$, and $(F_r(p), F_r(p'))$;

(ii) It is impossible to deduce s from $F_r(s)$ when r is unknown.

1.3. Self-distributive operations. Specifying an S-indexed family of mappings of a set S into itself amounts to specifying a binary operation on S, namely the operation $*$ defined by $x*y = F_x(y)$. Conversely, $(F_s)_{s\in S}$ is the family of all left translations for $(S, *)$. Now, in terms of the operation $*$, Condition (1.1) becomes

$$(1.2) \qquad r*(s*p) = (r*s)*(r*p),$$

i.e., it asserts that the operation $*$ satisfies the *left self-distributivity* law, usually denoted (LD) [6].

DEFINITION 1.1. A set equipped with a binary operation satisfying (1.2) is called an it LD-system.

Translating the previous authentication scheme into the language of LD-systems yields the following version.

Assume that $(S, *)$ is an LD-system. The public keys are a pair (p, p') of elements of S satisfying $p' = s*p$, while s is Alice's private key. The authentication procedure consists in repeating k times the following three exchanges:

A chooses r in S, and sends the commitments $x = r*p$ and $x' = r*p'$;
B chooses a random bit c and sends it to A;
For $c = 0$, A sends $y = r$, and B checks $x = y*p$ and $x' = y*p'$;
For $c = 1$, A sends $y = r*s$, and B checks $x' = y*x$.

---

2. LD-systems

The algebraic platforms eligible for implementing the scheme of Section 1 are LD-systems, and we are led to reviewing the existing examples of such algebraic systems.

2.1. Classical examples. A trivial example of an LD-system is given by an arbitrary set S equipped with the operation $x*y = y$, or, more generally,

$$x*y = f(y),$$

where f is any map of S into itself. Such examples are clearly not relevant for the scheme of Section 1, as the secret s plays no role in the computation.

The most classical example of an LD-system is provided by a group G equipped with the conjugacy operation

$$x*y = xyx^{-1}.$$

When G is a non-abelian group for which the conjugacy problem is sufficiently difficult, G is relevant for the scheme of Section 1, and, more generally, for the various schemes based on the Conjugacy Search Problem such as those of [21, 18] or [1]. Typical platform groups that have been much discussed in this context are Artin's braid groups B_n; in particular, the specific scheme considered in Section 1 is, in the case of the group B_n, (a variant of a scheme) proposed by H. Sibert in his PhD thesis [20].

2.2. The shifted conjugacy of braids.

Now, and this is the point we wish to emphasize here, examples of LD-system of a very different flavour exist.

Those LD-systems are connected with *free* LD-systems, *i.e.*, LD-systems that satisfy no other relations than those resulting from self-distributivity itself. It is easy to understand that a group equipped with conjugacy, even a free group, is not a free LD-system: indeed, the conjugacy operation always satisfies (among others) the idempotency law $x * x = x$, and the latter is not a consequence of (LD), as shows the existence of non-idempotent LD-system such as the integers equipped with $x * y = y + 1$.

Actually, free LD-systems are quite complicated objects, and we refer to [**6**], which contains an extensive description. For our purpose, it will be enough to know that, for some deep reasons that need not be explained here, there exists a simple self-distributive operation on Artin's braid group B_∞ that includes many copies of the free LD-system with one generator. Let us first recall the definition [**2, 6**]:

DEFINITION 2.1 (braid group). For $n \geqslant 2$, Artin's *braid group* B_n is defined to be the group with presentation

(2.1) $\quad \langle \sigma_1, ..., \sigma_{n-1} \,;\, \sigma_i \sigma_j = \sigma_j \sigma_i \text{ for } |i-j| \geqslant 2,\, \sigma_i \sigma_j \sigma_i = \sigma_j \sigma_i \sigma_j \text{ for } |i-j| = 1 \rangle.$

For each n, the identity mapping on $\{\sigma_1, ..., \sigma_{n-1}\}$ induces an embedding of B_n into B_{n+1}, so that the groups B_n naturally arrange into an inductive system of groups, and the limit is denoted by B_∞: this is just the group generated by an infinite family $\sigma_1, \sigma_2, ...$ subject to the relations (2.1).

LEMMA 2.2. *Let* d *be the shift mapping of the sequence* $(\sigma_1, \sigma_2, ...)$, *i.e., the function mapping* σ_i *to* σ_{i+1} *for each i. Then* d *induces an injective morphism of* B_∞ *into itself.*

PROOF (SKETCH). As the relations of (2.1) are invariant under shifting the indices, d induces a well-defined endomorphism of B_∞. That this endomorphism is injective follows from the interpretation of the elements of B_∞ in terms of braid diagrams [**2, 6**]: the geometric operation of deleting the leftmost strand is then well-defined, and it enables one to deduce $x = y$ from $\mathrm{d}x = \mathrm{d}y$. □

The main notion is then the following.

DEFINITION 2.3 (shifted conjugacy). For x, y in B_∞, we put

(2.2) $$x * y = x \cdot \mathrm{d}y \cdot \sigma_1 \cdot \mathrm{d}x^{-1}.$$

The above operation is a skew version of conjugation: y appears in the middle, and it is surrounded by x and x^{-1}; the difference with ordinary conjugation lies in the introduction of the shift d, and of the generator σ_1. The reader can check the equalities

$$1 * 1 = \sigma_1, \quad 1 * \sigma_1 = \sigma_2 \sigma_1, \quad \sigma_1 * 1 = \sigma_1^2 \sigma_2^{-1}, \quad \sigma_1 * \sigma_1 = \sigma_2 \sigma_1,$$

which show that shifted conjugation is quite different from conjugation.

PROPOSITION 2.4. [**4, 6**] *The system* $(B_\infty, *)$ *is an LD-system. Moreover, every braid generates under* $*$ *a free sub-LD-system.*

Checking that the operation defined in (2.2) satisfies the LD law is an easy verification. In the context of groups, the property that every element generates a free subgroup is torsion-freeness. Thus Proposition 2.4 expresses that $(B_\infty, *)$ is in some sense a torsion-free LD-system.

Understanding why the weird definition of shifted conjugacy has to appear requires a rather delicate analysis which is the main subject of the book [6]. It can be observed that, once the definition (2.2) is used, braids inevitably appear. Indeed, if we assume that G is a group, that f is an endomorphism of G, and that a is a fixed element of G, then defining

$$x * y = x\, f(y)\, a\, f(x)^{-1}$$

yields a left self-distributive operation (if and) only if the subgroup of G generated by the elements $f^n(a)$ is a homomorphic image of Artin's braid group B_∞, i.e., up to an isomorphism, it is B_∞ or a quotient of the latter group.

2.3. Discussion. Our intuition is that the LD-system $(B_\infty, *)$, i.e., braids equipped with shifted conjugacy, might be a promising platform for implementing the scheme of Section 1—or, more generally, for implementing any scheme based on a left self-distributive operation. This intuition ought to be confirmed by an experimental evidence, which at this early stage is not yet available. Here we content ourselves with a few remarks about the respective properties of conjugacy and shifted conjugacy in B_∞.

First, note that in general using free structures does not seem a very good idea in cryptography, as by definition the free structures are those in which the least possible number of equalities are satisfied, a not very good framework for hiding things. That is why, for instance, a free LD-system would probably not be the optimal platform for implementing the scheme of Section 1. However, the LD-system $(B_\infty, *)$ is far from being free, and it is even conjectured that it contains no free LD-system with two generators. For instance, the equality $\sigma_1 * \sigma_1 = \sigma_2 * \sigma_2$ ($= \sigma_2\sigma_1$) shows that the sub-LD-system generated by σ_1 and σ_2 is not free. No presentation of $(B_\infty, *)$ as an LD-system is known.

Practically, using shifted conjugacy of braids as suggested here relies on the difficulty of the following problem, which is analogous to the Conjugacy Search Problem:

> **Shifted Conjugacy Seach Problem**: Assuming that s, p are braids in B_∞ and $p' = s * p$ holds, find a braid $\widetilde{s}$ satisfying $p' = \widetilde{s} * p$.

Contrary to the Conjugacy Seach Problem, no solution to the Shifted Conjugacy Search Problem is known so far. It is not even known whether the simple Shifted Conjugacy Problem is decidable, i.e., whether one can effectively decide for two braids p, p' the existence of s satisfying $p' = s * p$. It is likely that shifted conjugacy is quite different from ordinary conjugacy, and that none of the many specific results established for the latter [14, 11, 15] extends to shifted conjugacy. In particular, we see no simple strategy for constructing the "shifted super summit set" of a braid p, defined as the family of all shifted conjugates of p with minimal canonical length—which is the key point in all solutions to the Conjugacy Problem known so far.

However, it is fair to mention that the Shifted Conjugacy Search Problem, which should not be threatened by specific attacks against the Conjugacy Search Problem [16, 19], remains, as the latter, an instance of the general Decomposition Problem and, as such, it is not a priori immune against length-based attacks [17, 12, 13].

To emphasize the difference between ordinary and shifted conjugations, we point

PROPOSITION 2.5 ([5], Corollary 1.8). *The mapping $f : s \mapsto s * 1$ is injective.*

In the case of ordinary conjugacy, every conjugate of 1 is 1, so the above injective function f is replaced with the constant function with value 1. By the way, very little is known about f. In particular, we raise

QUESTION 2.6. *Starting with a braid p, find s satisfying $s * 1 = p$ (when it exists).*

Once more, nothing is known. This might suggest to use f as a possible one-way function on braids.

3. The Laver tables and other algebraic systems

To conclude, we mention that braids are not the only possible platform for implementing self-distributive operations—and that the self-distributivity law is not even the only algebraic law eligible for the approach sketched in Section 1.

3.1. The Laver tables.
Instead of resorting to an infinite LD-system like B_∞ equipped with shifted conjugacy, one could instead use finite LD-systems. Such algebraic systems are far from being completely understood, but there exists an infinite sequence of so-called Laver tables that plays a fundamental role among LD-systems—similar to the role of the cyclic groups $\mathbb{Z}/p\mathbb{Z}$ among finite abelian groups—and, at the same time, has a high combinatorial complexity.

We refer to Chapter X of [6] for details. For our current overview, it is enough to mention that, for each nonnegative integer n, there exists a unique LD-system A_n such that the underlying set is the 2^n elements interval $\{0, 1, ..., 2^n - 1\}$ and one has $p * 0 = p + 1$ for $0 \leqslant p \leqslant 2^n - 2$ and $2^n - 1 * 0 = 0$. The value of $p * q$ in A_n can be easily computed using a double induction on q increasing from 0 to $2^n - 1$ and for p decreasing from $2^n - 1$ to 0, using the rule

$$p * (q + 1) = (p * q) * (p * 0),$$

and observing that $p * q$ has to be always strictly larger than p. Table 2 displays the first four Laver tables.

A_3	0	1	2	3	4	5	6	7
0	1	3	5	7	1	3	5	7
1	2	3	6	7	2	3	6	7
2	3	7	3	7	3	7	3	7
3	4	5	6	7	4	5	6	7
4	5	7	5	7	5	7	5	7
5	6	7	6	7	6	7	6	7
6	7	7	7	7	7	7	7	7
7	0	1	2	3	4	5	6	7

A_2	0	1	2	3
0	1	3	1	3
1	2	3	2	3
2	3	3	3	3
3	0	1	2	3

A_1	0	1
0	1	1
1	0	1

A_0	0
0	0

TABLE 1. The Laver tables A_n with $0 \leqslant n \leqslant 3$

Several general phenomena can be observed on these particular examples. First, for each n, the table A_n with 2^n elements is the projection modulo 2^n of the table A_{n+1} with 2^{n+1} elements. In other words, if we use a length n binary representation for the elements of A_n, only the dominant bit of each value has to be computed in order to determine A_{n+1} from A_n. Next, every row in the table A_n is periodic, with a period that is a power of 2. More precisely, for each p, the row of p in A_n consists of 2^k values

$$r_0 = p+1 < r_1 < ... < r_{2^k-1} = 2^n - 1$$

repeated 2^{n-k} times. One can show that, if $(r_0, ..., r_{2^k-1})$ is the periodic pattern in the row of p in A_n, with $r_0 = p+1$ and $r_{2^k-1} = 2^n - 1$ and if t denotes the smallest integer for which one has $p*t \geqslant 2^n$ in A_{n+1}, then
 - (i) either $t = 2^k$ holds, the period of p doubles from 2^k to 2^{k+1} between A_n and A_{n+1}, and the periodic pattern in A_{n+1} is $(r_0, ..., r_{2^k-1}, r_0+2^n, ..., r_{2^k-1}+2^n)$,
 - (ii) or $0 \leqslant t < 2^k$ holds, the period of p remains 2^k in A_{n+1}, and the periodic pattern in A_{n+1} is $(r_0, ..., r_{t-1}, r_t+2^n, ..., r_{2^k-1}+2^n)$.

In each case, the only piece of information needed to construct the row of p in A_{n+1} from the row of p in A_n is the value of t, which is called the *threshold* of p in A_n, and, therefore, the list of thresholds suffices to construct A_{n+1} from A_n (*cf.* Table 2). Note that, as A_n is the projection of A_{n+1}, we can consider that we work in the inverse limit A_∞ of the tables A_n, *i.e.*, we are constructing an LD-operation on 2-adic numbers.

A_0	1
	−

A_1	1	2
	0	1

A_2	1	2	3	4
	1	0	0	2

A_3	1	2	3	4	5	6	7	8
	2	2	1	0	0	0	0	4

TABLE 2. Threshold table for A_n with $1 \leqslant n \leqslant 3$

The reason for mentioning the Laver tables here is that their combinatorial properties seem to be very complicated. In particular, predicting the values in the first half of the sequence of thresholds is extremely difficult (the values in the second half are always $0, ..., 0, 2^n$): this is witnessed by the results of [8, 9, 10] which show that fast growing functions are necessarily involved here.

3.2. Central duplication. As a final remark, we come back to the Fiat–Shamir-like authentication scheme of Section 1. We noted that its security requires two conditions, namely one that is directly connected with the difficulty of what can be called the ∗-Search Problem, and the additional requirement that communicating $F_r(s)$, *i.e.*, $r*s$, gives no practical information about s when r remains unknown. Using the latter condition to forge an attack seems unclear, but, at least for aesthetic reasons, we might like to avoid it. This can be done, at the expense of changing the algebraic law.

Indeed, instead of communicating $F_r(s)$ in case $c = 1$ of the authentication scheme, Alice could communicate $F_s(r)$. In this case, the supposed difficulty of the ∗-Search Problem guarantees that $F_s(r)$ gives no information about s. Now, when the scheme is modified in this way, the equality checked by the verifier has to be modified as well. If we keep the same principle, we are led to replace Condition (1.1) with

(3.1) $$F_r(F_s(p)) = F_{F_s(r)}(F_r(p)).$$

When Condition (3.1) is translated into the language of binary operations, we obtain a new algebraic law, namely

$$(3.2) \qquad r * (s * p) = (s * r) * (r * p),$$

instead of left self-distributivity. Nothing specific is known about this law so far, but it should be possible to use the general method explained for a similar law in [7] to construct concrete examples of algebras that satisfy it.

4. Conclusion

We discussed various non-classical algebraic operations that could possibly be used as cryptographical primitives, typically for a Fiat–Shamir-like authentication scheme. The most promising example seems to be the shifted conjugacy operation on braids. At the least, the existence of such an operation shows that conjugacy is not the only possible primitive for braid-based cryptography, and that further investigation in this direction is needed.

References

[1] I. Anshel, M. Anshel, B. Fisher, & D. Goldfeld, *New key agreement protocols in braid group cryptography, CT–RSA 2001 (San Francisco, CA)*, Springer Lect. Notes Comp. Sci. **2020** (2001) 1–15.

[2] J. Birman, *Braids, Links, and Mapping Class Groups*, Annals of Math. Studies vol. 82, Princeton Univ. Press (1975).

[3] J.C. Cha, K.H. Ko, S.J. Lee, J.W. Han, J.H. Cheon, *An efficient implementation of braid groups, AsiaCrypt 2001*, Springer Lect. Notes in Comp. Sci. **2048** (2001) 144–156.

[4] P. Dehornoy, *Braid groups and left distributive operations*, Trans. Amer. Math. Soc. **345-1** (1994) 115–151.

[5] P. Dehornoy, *Strange questions about braids*, J. Knot Th. and its Ramifications **8-5** (1999) 589–620.

[6] P. Dehornoy, *Braids and Self-Distributivity*, Progress in Math. vol. 192, Birkhäuser (2000).

[7] P. Dehornoy, *Study of an identity*, Alg. Universalis **48** (2002) 223–248.

[8] R. Dougherty, *Critical points in an algebra of elementary embeddings*, Ann. P. Appl. Logic **65** (1993) 211–241.

[9] R. Dougherty & T. Jech, *Finite left-distributive algebras and embedding algebras*, Advances in Math. **130** (1997) 201–241.

[10] A. Drápal, *Persistence of left distributive algebras*, J. Pure Appl. Algebra **105** (1995) 137–165.

[11] N. Franco, J. González-Meneses, *Conjugacy problem for braid groups and Garside groups*, J. of Algebra **266-1** (2003) 112–132.

[12] D. Garber, S. Kaplan, M. Teicher, B. Tsaban, & U. Vishne, *Length-based conjugacy search in the braid group*, arXiv math.GT/0209267

[13] D. Garber, S. Kaplan, M. Teicher, B. Tsaban, & U. Vishne, *Probabilistic solutions of equations in the braid group*, Adv. Applied Math. **35** (2005) 323–334.

[14] F.A. Garside, *The braid group and other groups*, Quart. J. Math. Oxford **20-78** (1969) 235–254.

[15] V. Gebhardt, *A new approach to the conjugacy problem in Garside groups*, J. of Algebra **292–1** (2005) 282–302.

[16] D. Hofheinz & R. Steinwandt, *A practical attack on some braid group based cryptographic primitives*, PKC 2003; Springer Lect. Notes in Comp. Sci.; 2567; 2002; 187–198.

[17] J. Hughes & A. Tannenbaum, *Length-based attacks for certain group based encryption rewriting systems*, Workshop SECI02 Sécurité de la communication sur internet, September 2002, Tunis, http://www.storagetek.com/hughes/SECI02.pdf.

[18] K.H. Ko, S.J. Lee, J.H. Cheon, J.W. Han, J.S. Kang, & C. Park, *New public-key cryptosystem using braid groups, Crypto 2000*, Springer Lect. Notes Comp. Sci. **1880** (2000) 166–184.

[19] A.G. Myasnikov, V. Shpilrain & A. Ushakov, *A practical attack on some braid group based cryptographic protocols, Crypto 2005*, Springer Lect. Notes Comp. Sc. **3621** (2005) 86–96.

[20] H. Sibert, *Algorithmique des groupes de tresses*, Thèse de doctorat, Université de Caen (2003).

[21] V.M. Sidelnikov, M.A. Cherepnev, & V.Y. Yashcenko, *Systems of open distribution of keys on the basis of noncommutative semigroups*, Ross. Acad. Nauk Dokl. **332**-5 (1993) English translation: Russian Acad. Sci. Dokl. Math. 48-2 (1994) 384–386.

LABORATOIRE DE MATHÉMATIQUES NICOLAS ORESME UMR 6139, UNIVERSITÉ DE CAEN, 14032 CAEN, FRANCE

E-mail address: `dehornoy math.unicaen.fr`
URL: `//www.math.unicaen.fr/~dehornoy`

Length-based conjugacy search in the braid group

David Garber, Shmuel Kaplan, Mina Teicher, Boaz Tsaban, and Uzi Vishne

ABSTRACT. Several key agreement protocols are based on the following *Generalized Conjugacy Search Problem*: Find, given elements
$$b_1, \dots, b_n \text{ and } xb_1x^{-1}, \dots, xb_nx^{-1}$$
in a nonabelian group G, the conjugator x. In the case of subgroups of the braid group B_N, Hughes and Tannenbaum suggested a length-based approach to finding x. Since the introduction of this approach, its effectiveness and successfulness were debated.

We introduce several effective realizations of this approach. In particular, a length function is defined on B_N which possesses significantly better properties than the natural length associated to the Garside normal form. We give experimental results concerning the success probability of this approach, which suggest that an unfeasible computational power is required for this method to successfully solve the Generalized Conjugacy Search Problem when its parameters are as in existing protocols.

1. Introduction

Assume that G is a nonabelian group. The following problem has a long history and many applications (see [12]).

PROBLEM 1.1 (Generalized Conjugacy Search Problem). *Given elements $b_1, \dots, b_n \in G$ and their conjugations by an unknown element $x \in G$,*
$$xb_1x^{-1}, xb_2x^{-1}, \dots, xb_nx^{-1},$$
find x (or any element $\tilde{x} \in G$ such that $\tilde{x}b_i\tilde{x}^{-1} = xb_ix^{-1}$ for $i = 1, \dots, n$.)

In the sequel, we will not make any distinction between the actual conjugator x and any other conjugator $\tilde{x}$ yielding the same results.

This paper is a part of the Ph.D. thesis of the second named author at Bar-Ilan University.

This research was partially supported by the Israel Science Foundation through an equipment grant to the school of Computer Science in Tel-Aviv University. The authors were partially supported by: Golda Meir Fellowship (first named author), EU-network HPRN-CT-2009-00099(EAGER), ENRI, the Minerva Foundation, and ISF grant #8008/02-3 (second and third named authors), Koshland Fellowship (fourth named author).

©2006 American Mathematical Society

The *braid group* B_N is the group generated by the $N-1$ *Artin generators* $\sigma_1, \ldots, \sigma_{N-1}$, with the relations

$$\begin{aligned}\sigma_i \sigma_{i+1} \sigma_i &= \sigma_{i+1} \sigma_i \sigma_{i+1}, \\ \sigma_i \sigma_j &= \sigma_j \sigma_i \text{ when } |i-j| > 1\end{aligned}$$

Information on the basic algorithms in the braid group is available in [3] and the references therein. We will focus on the case where G is the subgroup of B_N generated by given elements $a_1, \ldots, a_m$. A solution of the generalized conjugacy problem in this case immediately implies the vulnerability of several cryptosystems introduced in [1, 15], and the methods of solution may be applicable to several other cryptosystems from [1, 18].

History, motivation, and related work. The length-based approach to the Conjugacy Problem was suggested by Hughes and Tannenbaum in [14], as a potential attack on the cryptosystems introduced in [1, 15]. Based on [14], Garrett [10] has doubted the security of these cryptosystems. But soon afterwards he published an errata withdrawing these doubts (see [12]). The reason was that no known realization of Hughes and Tannenbaum's scheme (i.e., definition of actual, effective length functions) was given before, and in particular, the success probability of this approach could not be estimated. The purpose of the current paper is to introduce and compare several such realizations, and provide actual success probabilities for specific parameters.

We stress that we are not interested here in the best possible solution of the generalized conjugacy problem, but rather in settling the debate concerning the applicability of the Hughes-Tannenbaum length-based approach to the problem.

Other approaches appear in [13, 17] and turn out more successful. However, the length-based approach has several advantages: First, one does not need to know the conjugated element in order to find the conjugator using this approach, and second, it essentially deals with *arbitrary equations*. The current paper gives the foundations of this approach, on which we build in [9], where an extension of this approach is suggested and good success rates are achieved for arbitrary equations.

Some of the citations of the present paper (see [2, 5, 11, 16, 19, 20, 21, 6]) refer to its preliminary draft [8], which contains much more details and examples. We have tried to make the present version concise.

Length-based attacks. Throughout this paper we make the following assumptions:

(1) The conjugator x belongs to a given finitely generated subgroup of B_N, whose generators

$$\{a_1, \ldots, a_m, a_1^{-1}, \ldots, a_m^{-1}\}$$

are given,

(2) x was generated as a product of a fixed, known number of generators $a_i^{\pm 1}$ chosen at random from the set of generators;
(3) We are given elements
$$xb_1x^{-1}, \ldots, xb_nx^{-1}$$
where each $b_i \in B_N$ is generated by some (nontrivial) random process, and we wish to find x.

We try to find the conjugator x by using the property that for an appropriate, efficiently computable *length function* ℓ defined on B_N, $\ell(a^{-1}ba)$ is usually greater than $\ell(b)$ for elements $a, b \in B_N$. Therefore, we try to reveal x by peeling off generator after generator from the given braid elements $xb_1x^{-1}, \ldots, xb_nx^{-1}$: Assume that
$$x = g_1 \cdot g_2 \cdots g_k,$$
where each g_i is a *generator*. We fix some *linear order* $\preccurlyeq$ on the set of all possible n-tuples of lengths, and choose a generator g for which the lengths vector
$$\langle \ell(g^{-1}xb_1x^{-1}g), \ldots, \ell(g^{-1}xb_nx^{-1}g) \rangle$$
is minimal with respect to $\preccurlyeq$. With some nontrivial probability, g is equal to g_1 (or at least, x can be rewritten as a product of k or fewer generators such that g is the first generator in this product), so that $g^{-1}x = g_2 \cdots g_k$ is a product of fewer generators and we may continue this way, until we get all g_i's forming x.

If one is capable of doing $O((2m)^t)$ computations, it is better to check all possibilities of $g_1 \cdots g_t$ by peeling off $g_1 \cdots g_t$ from x and choosing the t-tuple which yielded the minimal lengths vector. We will call this approach *look ahead of depth t*.

In order for any of the above to be meaningful, we must define the length function ℓ and the linear ordering $\preccurlyeq$. We will consider several candidates for these.

2. Realizations of the length function

We assume that each generator a_i is obtained by taking a product of some fixed number of (randomly chosen) Artin generators, to whom we refer as the "length" of the generators. Unless otherwise stated, in all of our experiments the length of each element a_i is 10. By a *generator* we mean either an element a_i, $i = 1, \ldots, m$, or its inverse. We will (informally) write $|x| = n$ when we mean that x was generated by a product of n generators chosen at random (with uniform distribution) from the list of $2m$ generators $a_1, \ldots, a_m, a_1^{-1}, \ldots, a_m^{-1}$.

2.1. The length function ℓ. The *Garside normal form* of an element $w \in B_N$ is the unique presentation of w in the form $\Delta_N^{-r} \cdot p_1 \cdots p_k$, where $r \geq 0$ is minimal and $p_1, \ldots, p_k$ are permutation braids in left canonical form [**3**]. Using the Garside normal form, one can assign a "length" to each $w \in B_N$ efficiently [**3**].

DEFINITION 2.1. The *Garside length* of an element $w \in B_N$, $\ell_G(w)$, is the number of Artin generators needed to write w in its Garside normal form. If the Garside normal form of w is $\Delta_N^{-r} \cdot p_1 \cdots p_k$, then

$$\ell_G(w) = r \cdot \binom{N}{2} + \sum_{i=1}^{k} |p_i|,$$

where $|p|$ denotes the length of the permutation p.[1]

The problem with this function is that it is not close enough to being monotone with $|x|$: One has to multiply many generators before an increase in the length function is observed. The left part of Figure 1 shows, for a fixed word b, $\ell_G(xbx^{-1})$ as a function of $|x|$. Its right part shows the *average* of $\ell_G(xbx^{-1})$ computed over 1200 random words.

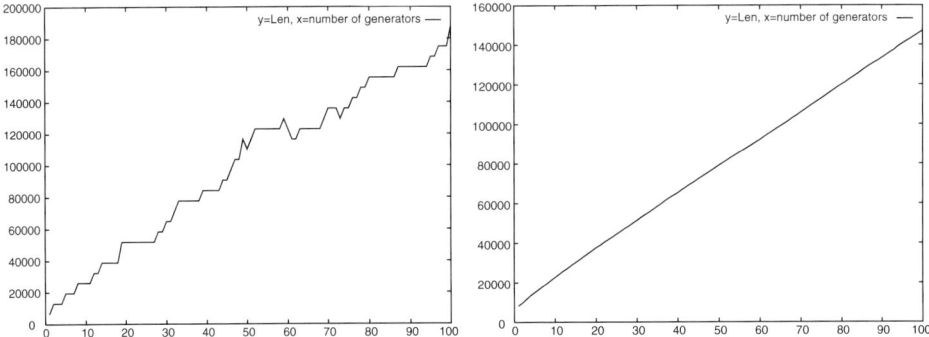

FIGURE 1. The growth of $\ell_G(w)$: Specific case (left) and average growth (right)

We wish to have a length function that is closer to being monotone. For each permutation braid p, $\tilde{p} := p^{-1}\Delta_N$ is a permutation braid. Thus, if $w = \Delta_N^{-r} \cdot p_1 \cdots p_k$ and $r > 0$, we can replace $\Delta_N^{-1} p_1$ with $\tilde{p}_1^{-1}$ to get $w = \Delta_N^{-(r-1)} \cdot \tilde{p}_1^{-1} p_2 \cdots p_k$. Now, Δ_N almost commutes with any permutation braid: For each permutation braid q there exists a permutation braid q' such that $|q'| = |q|$ and $q\Delta_N = \Delta_N q'$, that is, $\Delta_N^{-1} q^{-1} = (q')^{-1} \Delta_N^{-1}$. Consequently, $w = \Delta_N^{-(r-2)} \cdot (\tilde{p}_1')^{-1} \Delta_N^{-1} p_2 \cdots p_k$, and we can replace $\Delta_N^{-1} p_2$ with $\tilde{p}_2^{-1}$ as before. We iterate this process as much as possible, to get a presentation

$$w = \begin{cases} \Delta_N^{-(r-k)} (\tilde{p}_1')^{-1} \cdots (\tilde{p}_k')^{-1} & k < r \\ (\tilde{p}_1')^{-1} \cdots (\tilde{p}_r')^{-1} \cdot p_{r+1} \cdots p_k & r \leq k \end{cases}$$

---

[1]The length of a permutation p is the number of order distortions in p, that is, pairs (i,j) such that $i < j$ and $p(i) > p(j)$.

In each case, w has the form $a^{-1}b$ where a, b are positive braid words or the identity element, and we define the *reduced Garside length* to be the sum of the length of a and the length of b.[2] This is equivalent to the following.

DEFINITION 2.2. Let $w = \Delta_N^{-r} \cdot p_1 \cdots p_k$ be the Garside normal form of w. The *Reduced Garside length* of w is defined by

$$\ell_{\mathrm{RG}}(w) = \ell_{\mathrm{G}}(w) - 2 \sum_{i=1}^{\min(r,k)} |p_i|$$

This function turns out much closer to monotone than ℓ_{G} – see Figure 2.

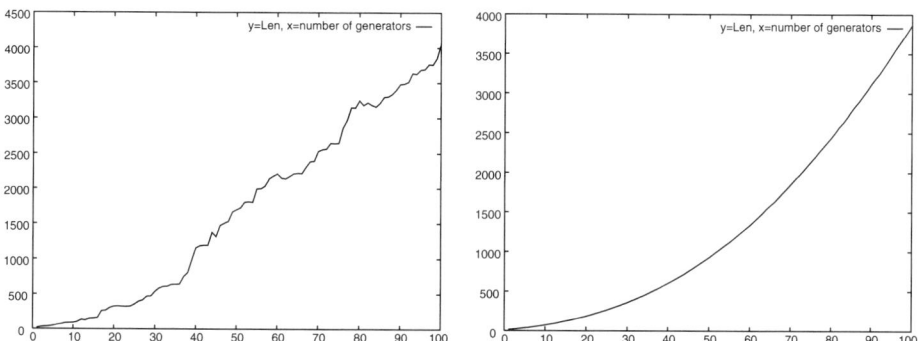

FIGURE 2. The growth of $\ell_{\mathrm{RG}}(w)$: Specific case (left) and average growth (right)

2.2. Statistical comparison of the length functions. The purpose of the length function ℓ is to distinguish between the case $|X| = k - 1$ (after peeling off a correct generator) and $|X| = k + 1$ (after trying to peel off a wrong generator). Thus, a natural measure for the effectiveness of the length function is the distance in standard deviations between $\ell(X')$ and $\ell(X)$ when $|X'| = |X| + 2$.

We fixed a random set of 20 generators in B_{81}, and computed (an approximation of) $E(\ell(X') - \ell(X))/\sqrt{V(\ell(X') - \ell(X))}$ as a function of $|X|$ for $|X| = 1, \ldots, 100$. (Roughly speaking, when n independent samples are added, the effectiveness of the comparison is $\sqrt{n}$ times this number.) We did that for both ℓ_{G} and ℓ_{RG}. The results appear in Figure 3, and show that the score for ℓ_{RG} is significantly higher. This phenomenon is typical – we have checked several random subgroups of the braid group and all of them exhibited the same behavior.

More evidence for the superiority of ℓ_{RG} over ℓ_{G} will be given in the following sections.

---

[2]The length of a positive braid word is well defined to be the number of generators in its presentation.

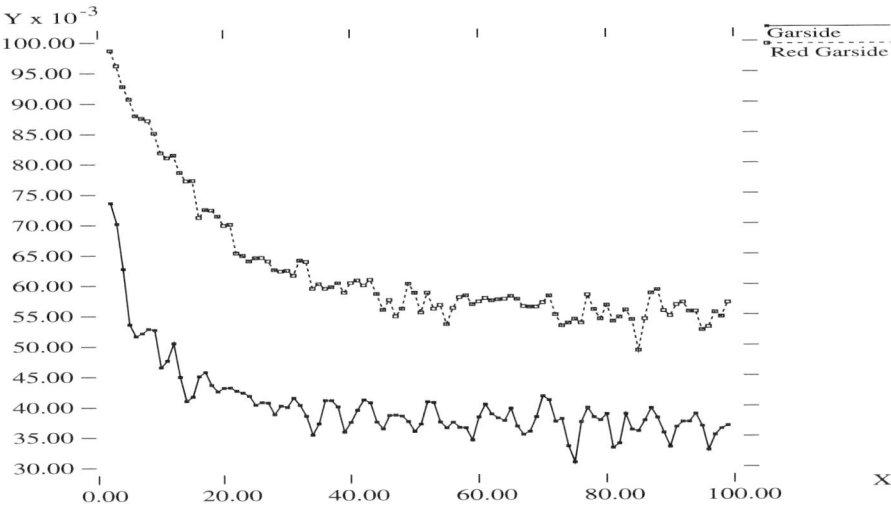

FIGURE 3. Distance between right and wrong in standard deviations.

3. Realizations of the linear ordering $\preccurlyeq$

Recall that after peeling off a candidate for a generator and evaluating the resulting lengths, we need to compare the vectors of lengths according to some linear ordering $\preccurlyeq$, and choose a generator which gave a minimal vector with respect to $\preccurlyeq$. We tested two natural linear orderings.

The most natural approach is to take the average of the lengths in the vector. This is equivalent to the following.

DEFINITION 3.1 (Average based linear ordering).

$$\langle \alpha_1, \ldots, \alpha_n \rangle \preccurlyeq_{\mathrm{Av}} \langle \beta_1, \ldots, \beta_n \rangle \quad \text{if} \quad \sum_{i=1}^{n} \alpha_i \leq \sum_{i=1}^{n} \beta_i.$$

With this at hand, we have performed the following experiment. We fixed a subgroup of B_{81} generated by $m = 20$ generators. Then we chose at random 200 elements of the form xw_j which share the same leading prefix x, and for each generator $a_i^{\pm 1}$ we computed $\ell(a_i^{\pm 1} xw_j)$ for each j (and $\ell = \ell_{\mathrm{G}}$ or ℓ_{RG}). For each of these two length functions ℓ, we have sorted the resulting length vectors according to $\preccurlyeq_{\mathrm{Av}}$ and checked the position of the "correct" generator, i.e., the generator which appeared leftmost in our computation of the word.[3] We repeated the computations for 138 distinct X's, and for $|X| = 40$ and $|X| = 100$. For an ideal length function (and an ideal linear ordering $\preccurlyeq$), the correct generator would always be ranked first, and the results in Figure 4 show that ℓ_{RG} is closer to this ideal than ℓ_{G}: In the graphs, we show the *distribution* (lower part of the graph) and

---

[3]In principle there could be more than one "correct" generator, but when the generators are long enough this is unlikely to happen often.

the *accumulated distribution* (upper part of the graph) of the position of the correct generator, for each of the length functions.

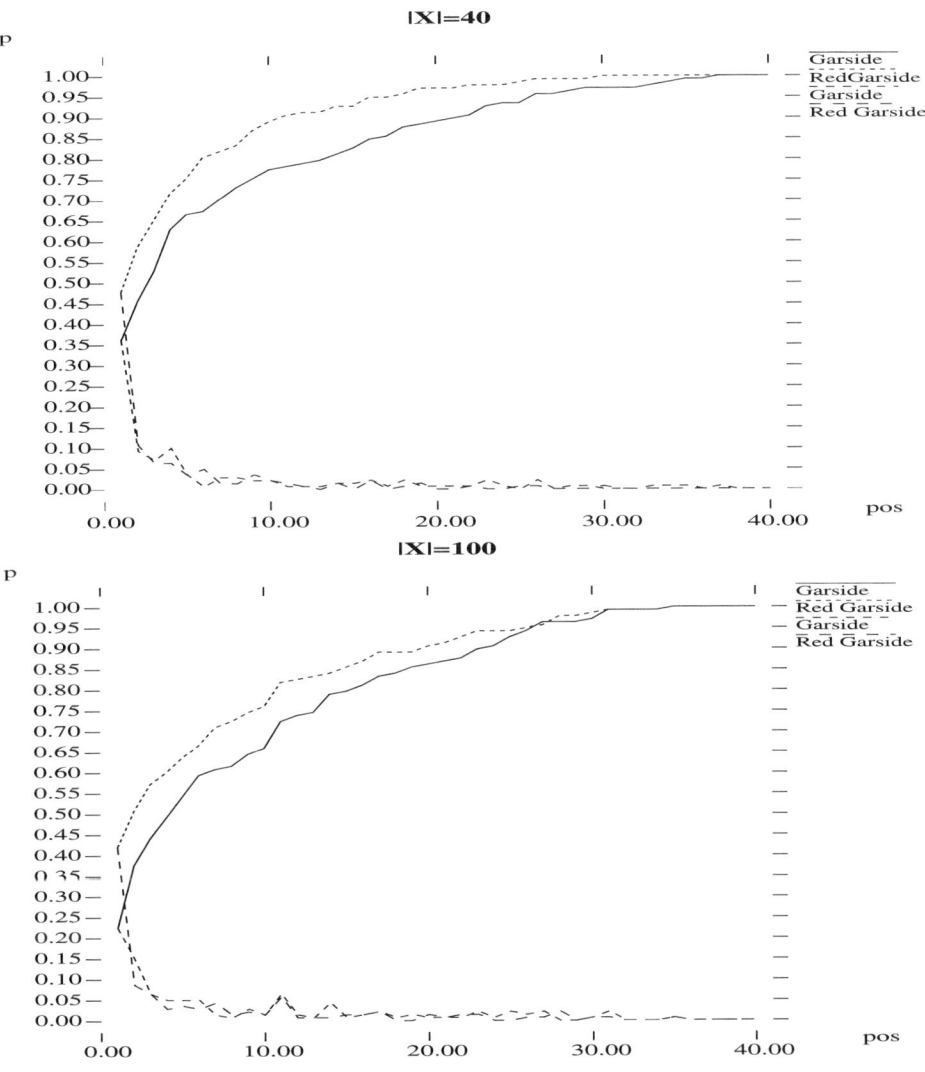

FIGURE 4. Position of correct generator ℓ_G and ℓ_{RG}

However, it turns out that even for the better length function ℓ_{RG}, the task of identifying the correct generator is not trivial. To demonstrate this, we selected at random one of the cases of x from the previous experiment, and computed over the given 200 samples the distribution of ℓ_{RG} for each generator. Figure 5 shows the distribution for the correct generator (in boldface) and of arbitrarily chosen 7 out of the remaining 40 generators (for an aesthetic reason we did not plot all 40).

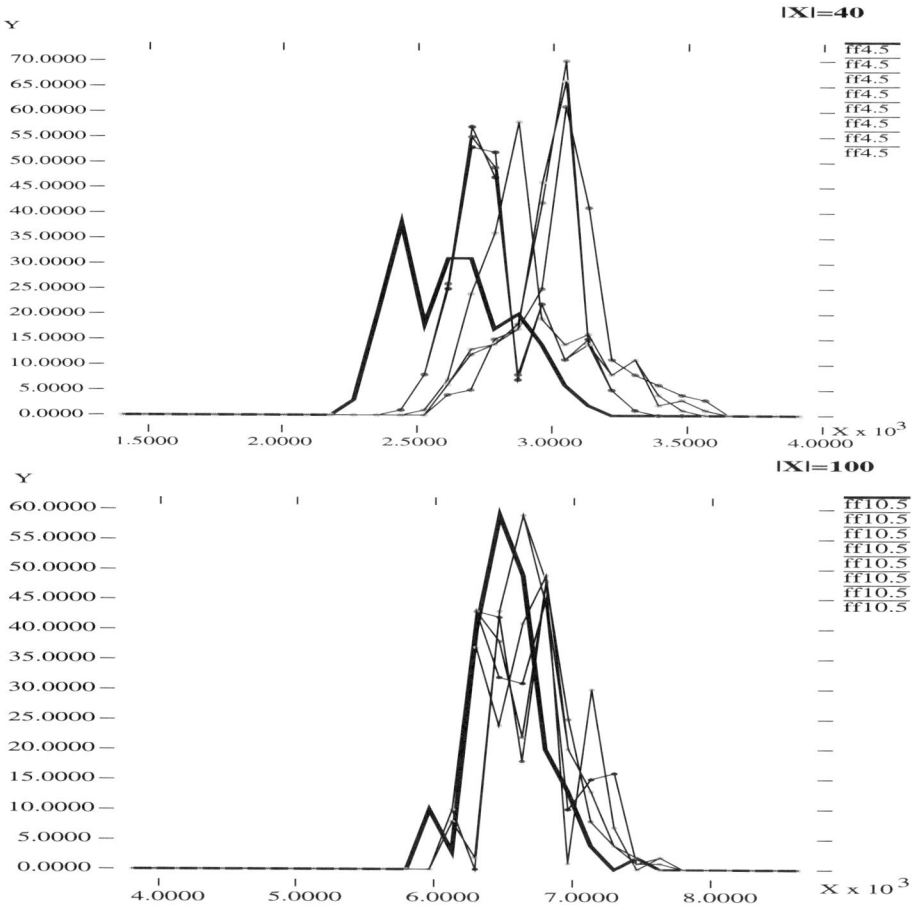

FIGURE 5. Actual distribution.

While the correct distribution tends more to the left (i.e., to smaller values), there is a large overlap with the rest of the distributions. We must emphasize that while Figure 5 demonstrates the typical case, there exist cases where the distribution of the correct generator is not the leftmost. In these cases the current method is doomed to fail, no matter how many conjugations we are given for the same conjugator.

Finally, for the sake of comparison, we define one more natural linear ordering of the space of length vectors. We expect the correct generator to yield the shortest length more often than the other generators. This motivates the following definition.

DEFINITION 3.2 (Majority based linear ordering). Consider the set of all obtained length vectors. For each $i = 1, \ldots, n$, consider the ith coordinate of each vector and let μ_i denote the minimum of all these ith coordinate values. Then

$$\langle \alpha_1, \ldots, \alpha_n \rangle \preccurlyeq_{\mathrm{Maj}} \langle \beta_1, \ldots, \beta_n \rangle \quad \text{if} \quad |\{i : \alpha_i = \mu_i\}| \geq |\{i : \beta_i = \mu_i\}|.$$

In the following section we compare the success probabilities of the length-based approach using the two length functions and two linear orderings defined in this section.

4. Experimental results for the conjugacy problem

4.1. The probability of obtaining the correct generator.
In this experiment we determine the probability that the correct generator is indeed the minimal with respect to the length function ℓ and linear ordering $\preccurlyeq$ used. The choice of parameters in the experiments throughout the paper are usually motivated by the choices given in [1], which are believed there to make the generalized conjugacy problem difficult.

We made 200 experiments using the following parameters: $N = 81$, n and m (the number of a_i's and b_i's, respectively) are both 20, the elements a_i and b_i are products of 10 random Artin generators, and x is the product of 5, 10, 20, 40, 60, or 100 random generators $a_i^{\pm 1}$, respectively. We tested look ahead depth $t = 1, 2$. In each cell of Table 1, below the probability that the correct generator is first, we wrote the probability of its being second.

	5	10	20	40	60	100
$\ell_G, \preccurlyeq_{Av}, t=1$	0.56	0.478	0.322	0.267	0.233	0.156
	0.16	0.188	0.1	0.167	0.089	0.1
$\ell_G, \preccurlyeq_{Maj}, t=1$	0.43	0.344	0.222	0.244	0.178	0.156
	0.14	0.178	0.144	0.122	0.1	0.044
$\ell_{RG}, \preccurlyeq_{Av}, t=1$	0.74	0.589	0.567	0.456	0.311	0.233
	0.13	0.233	0.189	0.122	0.167	0.167
$\ell_{RG}, \preccurlyeq_{Maj}, t=1$	0.71	0.578	0.578	0.433	0.289	0.211
	0.15	0.267	0.133	0.089	0.167	0.167
$\ell_G, \preccurlyeq_{Av}, t=2$	0.433	0.287	0.111	0.1	0.114	0.099
	0.156	0.08	0.087	0.038	0.055	0.035
$\ell_G, \preccurlyeq_{Maj}, t=2$	0.25	0.147	0.103	0.058	0.086	0.03
	0.033	0.036	0.024	0.008	0.023	0.03
$\ell_{RG}, \preccurlyeq_{Av}, t=2$	0.578	0.526	0.333	0.242	0.2	0.168
	0.183	0.127	0.135	0.138	0.105	0.05
$\ell_{RG}, \preccurlyeq_{Maj}, t=2$	0.511	0.482	0.31	0.242	0.186	0.149
	0.139	0.139	0.127	0.104	0.091	0.089

TABLE 1. The probability that the correct generator is first or second

Table 1 shows that the Reduced Garside length function ℓ_{RG} is significantly better than the standard Garside length function ℓ_G. Also, observe that using look ahead depth 2 is preferable to using look ahead depth 1 twice (to see this, square the probabilities for $t = 1$). Another natural approach to using look ahead $t > 1$ is to consider only the first generator (of the word with the least score) as correct, and ignore the rest of the generators. This means that in the algorithm for finding x, we peel off only one generator at a time despite the fact that we used look ahead $t > 1$. This gives better success rates than just taking $t = 1$, and our experiments indicate that this approach may be slightly better than that of taking the whole look ahead word, but we did not extensively check this conjecture since the differences were not significant. Some other variants of the usage of look ahead are mentioned in [**8**].

4.2. Nonsymmetric parameters.
This experiment checks the effect on the probability of success when the lengths of the generators a_i and elements b_i (in terms of Artin generators) are not equal.

We tested the probability of success for $N = 81$, $n = m = 20$, look ahead depth $t = 2$, and $|x| = 30$.

	length of b_i	ℓ_{RG} a_i of length:					ℓ_{G} a_i of length:				
		5	10	15	20	25	5	10	15	20	25
$\preccurlyeq_{\text{Av}}$	5	44	82	124	134	156	32	51	81	102	115
	10	59	97	113	141	150	56	69	79	91	96
	15	56	91	123	136	141	31	53	75	93	105
	20	49	77	115	132	149	31	49	77	86	107
	25	56	84	102	127	141	42	59	60	91	100
$\preccurlyeq_{\text{Maj}}$	5	39	70	121	134	160	28	41	66	84	87
	10	57	97	114	140	156	49	49	58	83	82
	15	50	85	118	136	144	19	45	59	73	89
	20	48	80	116	133	149	39	41	52	73	86
	25	60	89	101	141	152	39	50	56	72	78

TABLE 2. Number of success out of 200 tries for different lengths

As expected, Table 2 shows that if the length of the elements a_i increases then so does the probability to find a correct generator (this is like making look ahead deeper without exponentially increasing the number of candidates for the prefix of x). On the other hand, the effect of the length of the elements b_i is not significant.

4.3. Increasing the number of given conjugates.
Several experiments showed that increasing the number n of given elements xb_ix^{-1} from few (about 10) to many (about 3000) did not significantly increase the probability that the correct generator appears first.

In an instance of the problem the length function ℓ and the (unknown) element x are fixed, and this defines for each generator g the distribution F_g of $\ell(g^{-1}xbx^{-1}g)$ over random words b of a fixed given length (in terms of Artin generators). For each g, we have a sample of the distribution F_g for each given equation. In most cases, the expectancy of F_g where g is the first letter in x is smaller than the other expectancies (see Section 3), and then enough samples will allow us to identify g. However in some cases the minimal expectancy is obtained for another generator. In these cases adding more samples cannot help, and so the probability to find the correct generator does not tend to 1 when we increase the number of samples.

Another important observation is that few samples (about 15) are needed in order to get very close to the expectancy of the distributions F_g. In light of the preceding paragraph, the outcome of the algorithm can be decided after a relatively small number of samples (i.e., given conjugates) are collected. In particular, the success probability does not significantly improve when n is large.

4.4. Finding x.

The simplest way to try and obtain all generators of x and therefore x would be to use any of the above algorithms iteratively, at each step peeling off the first generator. In the following experiment, the probability to find *all of* x this way was tested. Here too, the lengths of the a_i's and b_i's were 10 Artin generators. We made 500 experiments, using a weaker variant of ℓ_{RG} as the length function, and with no look ahead ($t=1$). We repeated this for $B_4, \cdots, B_{20}$ and x of lengths 2 to 18 generators $a_i^{\pm 1}$. The result is the number of successes out of 500 tries.

N	2	3	4	5	6	7	8	9	10	11	12	13	14	15	16	17	18
4	429	361	289	262	204	181	137	120	107	94	77	52	50	37	38	25	31
5	436	378	327	269	215	185	173	120	119	106	75	67	56		44		28
6	446		324	282	243		183	154	115	107	88	68	65	59	36	47	
7	453	400	330	287		208	176	142	126	97	74	69	50		35	39	33
8	440	396	275	230	198	149	137	116	103	63	57	51	39	34	37	25	23
9	463	404	334	276	208	180	148	121	86	70	73	41	44	29	29	17	15
10	461	383	328	274	221	165	156	113	83	71	60	46	42	30	26	10	17
14	460	377	295	244		140	108	79	54	41	33	19	14	14	8	9	8
17	453	365	293	221	167	118	89	56	56	33	16	17	10	4	2	4	
20	455	373	305	226	153		73	43	36	21	11		8		3	3	2

TABLE 3. Number of successes for finding x out of 500 tries

The results suggest that while we already obtain solutions (with nontrivial probability) for some nontrivial parameters, we must extend the approach in order to consider harder parameters. A successful extension is discussed in [9]. In the sequel we discuss some other possible extensions.

5. Possible improvements and conclusions

One approach is to create new conjugates by multiplying any number of existing ones (or their inverses). In fact, if $\mathcal{B}$ is the group generated by $b_1, \ldots, b_n$, then the group generated by $xb_1x^{-1}, \ldots, xb_nx^{-1}$ is $x\mathcal{B}x^{-1}$. By Section 4.3, this does not help much.

The algorithm can be randomized by conjugating the given elements $xb_1x^{-1}, \ldots, xb_nx^{-1}$ by a random (known) element $y \in \langle a_1, \ldots, a_m \rangle$, so that running it several times increases the success probability. The problem with this approach is that the conjugator becomes longer and therefore the probability of success in each single case decreases.

Our experiments showed that the peeling off process often enters a loop, that is, a stage to which we return every several steps. This can sometimes be solved by conjugating with a random known element after we enter the loop. We also tried to change the length function or the linear ordering when we enter a loop. These approaches were successful for small parameters but did not result in a significant improvement for large parameters.

We did not try approaches of learning algorithms, neural networks, etc. A simple example is to try and learn the distribution of the lengths for the correct generator and define the linear ordering according to the likelihood test.

The purpose of this paper was to check the applicability of Hughes and Tannenbaum's length-based approach against the key agreement protocols introduced in [1, 15]. Our results suggest that this approach requires an unfeasible computational power in order to solve the generalized conjugacy search problem for the parameters used in these protocols. However, this method has natural extensions which can make it applicable: In [9] we suggest one particularly successful extension, and it turns out that it can solve these and other problems with standard computational power.

References

[1] I. Anshel, M. Anshel and D. Goldfeld, *An algebraic method for public-key cryptography*, Mathematical Research Letters **6** (1999), 287–291.
[2] P. D. Bangert, *Raid Braid: Fast Conjugacy Disassembly in Braid and Other Groups*, Applications of Computer Algebra (ACA-2004).
[3] J. C. Cha, K. H. Ko, S. J. Lee, J. W. Han and J. H. Cheon, *An Efficient Implementation of Braid Groups*, ASIACRYPT 2001, Lecture Notes in Computer Science **2248** (2001), 144–156.
[4] P. Dehornoy, *A fast method for comparing braids*, Advances in Mathematics **125** (1997), 200–235.
[5] P. Dehornoy, *Braid-based cryptography*, in: **Group Theory, Statistics, and Cryptography**, Contemporary Mathematics **360** (2004), 5–33.
[6] P. Dehornoy, *Using shifted conjugacy in braid-based cryptography*, Contemporary Mathematics, to appear.
[7] D. B. A. Epstein, M. S. Paterson, G. W. Camon, D. F. Holt, S. V. Levy, and W. P. Thurston, **Word Processing in Groups**, Jones and Bartlett, Boston, MA, 1992.
[8] D. Garber, S. Kaplan, M. Teicher, B. Tsaban, and U. Vishne, *Length-based conjugacy search in the braid group*, Version 1, Mathematics ArXiv eprint (2002) `math.GR/0209267 (v.1)`
[9] D. Garber, S. Kaplan, M. Teicher, B. Tsaban, and U. Vishne, *Probabilistic solutions of equations in the braid group*, Advances in Applied Mathematics **35** (2005), 323–334.
[10] P. Garrett, *Making, Braking Codes: Introduction to Cryptology*, Prentice Hall, 2000.
[11] J. Gozález-Meneses, *Improving an algorithm to solve multiple simultaneous conjugacy problems in braid groups*, in: **Geometric methods in group theory**, Contemporary Mathematics **372** (2005), 35–42.
[12] Helger Lipmaa, *Cryptography and Braid Groups homepage*, `http://www.tcs.hut.fi/~helger/crypto/link/public/braid/`

[13] D. Hofheinz and R. Steinwandt, *A Practical Attack on Some Braid Group Based Cryptographic Primitives*, Proceedings of PKC 2003, Lecture Notes in Computer Science **2567** (2003), 187–198.

[14] J. Hughes and A. Tannenbaum, *Length-based attacks for certain group based encryption rewriting systems*, Workshop SECI02 Sécurité de la Communication sur Internet, September 2002.

[15] K. H. Ko, S. J. Lee, J. H. Cheon, J. W. Han, S. J. Kang and C. S. Park, *New Public-key Cryptosystem using Braid Groups*, CRYPTO 2000, Lecture Notes in Computer Science **1880** (2000), 166–183.

[16] E. Lee, *Braid groups in cryptology*, IEICE Transactions on Fundamentals of Electronics, Communications and Computer Sciences **E87-A** (2004), 986–992.

[17] E. Lee, J. H. Park, *Cryptanalysis of the public-key encryption based on braid groups*, EUROCRYPT 2003, Lecture Notes in Computer Science **2656** (2003), 477–490.

[18] S. H. Paeng, K. C. Ha, J. H. Kim, S. Chee and C. Park, *New Public Key Cryptosystem Using Finite Non Abelian Groups*, CRYPTO 2001, Lecture Notes in Computer Science **2139** (2001), 470–485.

[19] V. Shpilrain, *Assessing security of some group based cryptosystems*, Contemporary Mathematics **360** (2004), 167–177.

[20] A. Ushakov and V. Shpilrain, *The conjugacy search problem in public key cryptography: unnecessary and insufficient*, Applicable Algebra in Engineering, Communication and Computing, to appear.

[21] V. Shpilrain and G. Zapata, *Combinatorial group theory and public key cryptography*, Applicable Algebra in Engineering, Communication and Computing, to appear.

DAVID GARBER, EINSTEIN INSTITUTE OF MATHEMATICS, THE HEBREW UNIVERSITY, GIVAT-RAM 91904, JERUSALEM, ISRAEL; AND DEPARTMENT OF SCIENCES, HOLON ACADEMIC INSTITUTE OF TECHNOLOGY, 52 GOLOMB STREET, HOLON 58102, ISRAEL
E-mail address: `garber@math.huji.ac.il, garber@hait.ac.il`

SHMUEL KAPLAN, MINA TEICHER, AND UZI VISHNE, DEPARTMENT OF MATHEMATICS AND STATISTICS, BAR-ILAN UNIVERSITY, RAMAT-GAN 52900, ISRAEL
E-mail address: `[kaplansh, teicher, vishne]@math.biu.ac.il`

BOAZ TSABAN, DEPARTMENT OF MATHEMATICS, THE WEIZMANN INSTITUTE OF SCIENCE, REHOVOT 76100, ISRAEL
E-mail address: `boaz.tsaban@weizmann.ac.il`
URL: `http://www.cs.biu.ac.il/~tsaban`

Towards Provable Security for Cryptographic Constructions Arising from Combinatorial Group Theory

María Isabel González Vasco, Rainer Steinwandt, and Jorge L. Villar

ABSTRACT. More than two decades after the stipulation of *semantic security* by Goldwasser and Micali, the framework of provable security has become a widely accepted tool for exploring the security of cryptographic schemes. The established models allow to deal with a significant class of attacks and proved their practical relevance in the design and analysis of actually deployed schemes.

In this contribution, we address the impact of modern security analysis in the sense of provable security to cryptographic proposals building on combinatorial group theory. Using security deficiencies in proposals of Wagner and Magyarik and of Birget et al. as examples, we reiterate on the need for carrying out a rigorous cryptographic security analysis when trying to base cryptographic schemes on a new mathematical platform. To this aim, we suggest possible guidelines that target at facilitating the development of modern asymmetric encryption schemes building on non-abelian groups.

1. Introduction

Since the early days of public key cryptography, it was understood that a purely intuitive handling of security requirements for cryptographic tools is not sufficient to prevent undesirable situations in practice. To deal with this problem, formal frameworks have been developed trying to specify desired security requirements in a manner that is both amenable to mathematical proofs and meaningful for applications. Certainly the available tools are not impeccable and typically do, e.g., not protect against attacks on the implementation level. Nevertheless, the available frameworks have turned out to be of great value in the design of robust cryptographic schemes. For more detailed descriptions of the theory of provable security and its historical motivation see, for instance, [**Ste03, Poi04, KMar**].

For public key encryption schemes the (by now standard) notion of IND-CCA2 security is particularly satisfying in the sense that other security characterizations like NM-CCA2 (see, e. g., [**BDPR98**]) and a functionality in the UC-framework (cf., e. g., [**CKN03, HMQSar**]) yield equivalent security requirements. In this

---

1991 *Mathematics Subject Classification.* Primary 94A60; Secondary 20F10.

Key words and phrases. provable security, combinatorial group theory, cryptography.

Work of second author partially supported by a Federal Earmark grant for *Research in Secure Telecommunication Networks (2005-06)*.

contribution we focus on the design of public key encryption schemes, and, in particular, on recent proposals built on combinatorial group theory.

Indeed, combinatorial group theory provides a source of well explored computational problems that could prove valuable for cryptographic applications. However, it is often the case that proposals of schemes arising in this area lack a formal security analysis in modern terms and succumb to cryptographically very simple attacks. In the sequel we show different examples of this situation, and suggest possible design guidelines that hopefully facilitate the design of more robust future proposals.

In more detail, our contribution is organized as follows: Section 2 gives a short summary of the *provable security* framework for encryption schemes. Subsequently, we present in Section 3 several examples of schemes basing on combinatorial group theory that have been proven insecure in modern terms; that is, simple attacks taken into account by modern security models are successful (and these attacks do not invalidate the soundness of the mathematical assumptions considered). Section 4 is devoted to a non-abelian variant of Cramer and Shoup's framework for designing provable secure encryption schemes in the stricter sense; through this framework, more sound cryptographic constructions exploiting the hardness of some combinatorial group theory problems might be deployed: Once certain mathematical assumptions are met, the outlined construction guarantees security in the sense of IND-CCA2. We conclude this contribution with some final remarks.

2. Provable security for encryption schemes (in brief)

It is established cryptographic practice to propose a new encryption scheme along with a concise security model clearly specifying the assumptions made and security guarantees provided. Besides helping to avoid misunderstandings, being precise about assumptions and guarantees also facilitates the design and analysis of more complex protocols that use an encryption scheme as a building block. Subsequently we shortly recall some basic security notions for encryption schemes at an informal level. More detailed and formal definitions can be found in [**Ste03, Poi04, KMar**], for instance. An important observation to make is that the security notions do not refer to a specific mathematical assumption. Only the security guarantees to be offered by the resulting encryption scheme are specified. The choice of the underlying mathematical hardness assumption that is used to implement the security guarantee is left to the system designer.

The idea of a "feasible" computation is usually captured as being computable in (probabilistic) polynomial time. In particular, a public-key encryption scheme is usually defined as a collection of three feasible algorithms in this sense:

DEFINITION 2.1. *A public key encryption scheme consists of three polynomial time algorithms* $(\mathcal{K}, \mathcal{E}, \mathcal{D})$ *such that:*

$\mathcal{K}$: *the* key generation algorithm *is probabilistic, and on input a security parameter* $\ell \in \mathbb{N}$ *outputs a pair consisting of a private and a public key* (sk, pk);

$\mathcal{E}$: *the* encryption algorithm *is probabilistic, and receives as an input the public key and a message* m. *It outputs a ciphertext* c, *which can be denoted as* $\mathcal{E}(pk, m, r)$ *where* r *stands for the random value(s) used by* $\mathcal{E}$;

$\mathcal{D}$: *the* decryption algorithm *is deterministic, and on input the secret key sk and a ciphertext c outputs either the plaintext associated to c or an error message (if c is not a valid ciphertext according to $\mathcal{E}$ and pk).*

All messages and keys above may be interpreted as bitstrings of length bounded by a (polynomial) function in the security parameter ℓ.

Following the same intuition, feasible adversaries are often modeled as probabilistic polynomial time (ppt) algorithms. A typical security proof gives a reduction argument showing that the existence of any ppt algorithm violating a certain security goal contradicts a mathematical assumption. For common security goals, an adversary is considered as successful, if the probability of a successful attack is *non-negligible*, which can be taken for a "polynomial probability of success". Adversarial goals commonly discussed are the following:

Violating the indistinguishability (IND) of encryptions: informally, we say the adversary wins if he can derive any bit of information about the plaintext (except its length) seeing only the corresponding ciphertext. An encryption scheme resistant to this kind of attacks is called *semantically secure*. The notion of semantic security was introduced by Goldwasser and Micali [**GM84**]. Note that this requirement is significantly stronger than "hiding the complete plaintext". For instance, semantic security does not tolerate the existence of a feasible adversary that can recover the least significant bit of the plaintext underlying encountered ciphertexts.

In most settings, semantic security is equivalent to an easier-to-prove security notion called *indistinguishability of encryptions*. To prove an encryption scheme secure in the latter sense, one considers the following game that also explains the name of this security notion: the adversary is presented with the public key pk output by $\mathcal{K}$, and chooses two different plaintext messages m_0 and m_1 of equal length. Subsequently, the adversary is presented with the encryption of one of these messages (using $\mathcal{E}$ and pk), his *target ciphertext*, and must guess which of the two plaintexts of his choice corresponds to it. Then, if for any feasible adversary the probability of success is at most negligibly away from $1/2$ ("guessing"), the scheme achieves indistinguishability of encryptions.

The game just described captures a situation where the set of possible plaintexts is not only amenable to an exhaustive search, but even determined by the adversary. Even in this situation, the secrecy of the message is to be ensured with overwhelming probability. As the adversary knows the public encryption key pk and the encryption algorithm $\mathcal{E}$, in particular no deterministic scheme can guarantee indistinguishability of encryptions.

Violating the non-malleability (NM) of encryptions: informally, the adversarial goal is to construct, given a ciphertext c_0 (which now constitutes the adversary's *target*), new ciphertexts $c_1, \ldots, c_k$ such that the plaintexts underlying $c_1, \ldots, c_k$ are "meaningfully related" to the plaintext underlying the target. This notion is due to Dolev, Dwork and Naor, and more details can be found in [**DDN00, BDPR98**]. A typical example motivating malleability is a setting where two competing parties are to send an encrypted offer for some product or service over a public channel. In a malleable scheme, it could be possible that eavesdropping an encrypted offer enables a competitor to compute a ciphertext encoding a better offer. The important observation to make is that a successful attack against non-malleability does not impose the adversary to learn parts of the encrypted plaintext. An attack is already successful if *ciphertexts* can be found such that after application of the decryption algorithm a known relation among the resulting plaintexts holds.

For a public-key encryption scheme it is natural to assume that an adversary can access the public key and therewith encrypt any message of his choice. In addition this, common security models also assume that the adversary has rather generous black box access to a an oracle decrypting ciphertexts by running the algorithm $\mathcal{D}$. The expository article [**Sho98**] gives a good idea of "why chosen ciphertext security matters". It is by now common practice to grant the adversary access to a decryption oracle with the sole restriction being that the target ciphertext itself must not be sent to the oracle. In particular, the adversary is allowed to send invalid ciphertexts. This is crucial, as error messages produced by a decryption algorithm may reveal critical information. Depending on the application contexts, weaker adversarial models may be justified, but also in the design of encryption schemes such weaker models have their justification: A security proof for a weaker adversarial model might be significantly easier to achieve, and then a generic construction can be applied to "lift" a scheme to a higher security level. A selection of typically considered adversarial capabilities is the following:

CPA adversaries: they only have access to the encryption algorithm, and knowledge of all public parameters and keys. That is, they are in the weakest possible situation for a public-key encryption scheme and are restricted to *chosen plaintext attacks*.

VCA adversaries: besides the public information and encryption algorithm they have access to a *validity checking oracle* which allows them to identify properly formed ciphertexts (see [**HGS99**]). This corresponds to a situation where, say, a server either accepts received ciphertexts or rejects them with an error message giving no further information about the reason for a decryption failure.

CCA1 adversaries: in addition to what a CPA adversary has, they have limited (non-adaptive) access to a decryption oracle. That is, the adversary can have arbitrary messages decrypted *before* he gets to know his challenge ciphertext. This type of chosen ciphertext attack is sometimes referred to as *lunch time attack*: The adversary has temporary (lunch time) access to the decryption algorithm, but the challenge ciphertext is received "after lunch".

CCA2 adversaries: like CCA1 adversaries, they have limited access to a *decryption oracle*, and can feed it with anything different than the challenge ciphertext itself. Differing from the CCA1 setting, access to the decryption oracle remains possible after the challenge ciphertext is known. Thus, queries to the decryption oracle can be adaptive in a very strong sense.

Security notions are defined by combining a security goal with an assumption on the adversarial capabilities: IND-VCA, NM-CPA, etc. Further details and a good survey on how various of these notions are related can be found in [**BDPR98**], for instance. The strongest security notion typically considered is IND-CCA2, which has been proven equivalent to NM-CCA2[1]. Unfortunately, very few schemes allow for a proof of security in this very strict sense in the standard model, that is without any extra assumptions. An example of such a construction is the celebrated Cramer-Shoup methodology [**CS02**], which yields a construction of IND-CCA2 secure schemes from a new mathematical primitive called *hash proof systems*. Their construction can actually be understood in group theoretical terms, and generalized in order to allow for the use of very general algebraic structures. We will briefly review this work in Section 4.

3. Security of group based proposals

Combinatorial group theory has, since the early days of public key cryptography, appeared as a potential source of computationally hard problems. Since the late nineties, much attention has been attracted by this area of research, in particular due to the proposals based on braid groups [**AAG99, KLC$^+$00**]. For a detailed and comprehensive survey on braid group based cryptography we refer to [**Deh04**].

Roughly speaking, many braid group based proposals are based in problems related to the conjugacy problem, and mimic discrete logarithm based constructions. It is worth noting that some of the braid based proposals have been analyzed in modern terms, that is, according to *provable security* frameworks as mentioned above. Therefore, the security flaws identified in them are (generally speaking) not a consequence of naive cryptographic design, but rather of unlucky choice of the underlying mathematical assumptions. Consequently, here the cryptanalytic success can contribute to a better understanding of the underlying mathematical problem instead of "only breaking a cryptographic scheme".

As exemplified in the sequel, several other proposals originating in combinatorial group theory are less satisfying in the sense, that from a cryptographic point of view they can be broken in a rather simple manner: exhibited vulnerabilities are independent of the soundness of the mathematical assumptions considered. Prominent examples for this situation are word/rewriting based schemes.

3.1. Reaction attacks on word problem based schemes. Already at CRYPTO '84 Wagner and Magyarik [**WM85**] gave some guidelines on how hardness of the word problem in finitely presented groups could be applied to the design of public key encryption schemes. This problem can be stated as follows: given a finite presentation of a group (X, R) decide wether two words over $X \cup X^{-1}$ represent the same group element. Subsequent to this seminal work and up to present, there have been several proposals for deriving encryption schemes using related rewriting problems in a similar fashion (sometimes in weaker algebraic structures), see, for instance, [**ATS03, GZ91, SM90, STAS02**]. However, in most cases the security arguments supporting these schemes were rather heuristic, and the schemes were subsequently proved to be vulnerable to different types of attacks (cf. [**GVS01, GVS04, Pet03, GVSar, GVHMS04, dVP04, Per05**]). To illustrate this situation, let us briefly describe how the VPA/reaction attack from [**GVS04**] works:

---

[1] It is actually quite standard to omit the number "2" from CCA2 and use the term CCA when referring to this notion.

3.1.1. *Wagner and Magyarik's proposal.* Let G be a group defined by the finite presentation (X, R), for which the word problem is hard to solve. Moreover, suppose S is a set of words in $(X \cup X^{-1})^*$ such that for the quotient group $\widetilde{G}$, specified by the presentation $(X, R \cup S)$, there exists a polynomial time algorithm $\mathcal{A}$ for solving the word problem.

Let Σ be a finite alphabet and $W(\Sigma) = \{ w_\sigma \mid \sigma \in \Sigma \}$ a subset of $\{X \cup X^{-1}\}^*$ such that if $\sigma \neq \tau$, then w_σ and w_τ are neither equivalent over G nor over $\widetilde{G}$. While the presentation (X, R) and the set $W(\Sigma)$ are made public, the set S is kept secret. To encrypt an element $\sigma \in \Sigma$, the scheme of Wagner and Magyarik proceeds as follows: take w_σ and rewrite it using the public relations specified by R. The resulting word w is the ciphertext. To decrypt, a recipient runs the algorithm $\mathcal{A}$ with inputs $w w_\sigma^{-1}$ ($\sigma \in \Sigma$).

For more details on how to choose the parameters of the above scheme, we refer to [**WM85**]. This scheme was supposed to be "secure" provided that:

- the word problem in G is hard enough, and
- for most quotient groups of G with easy word problem all the words in $W(\Sigma)$ should be equivalent to the empty word (that is, constructing an "alternative" secret key should not be possible).

3.1.2. *A VPA/Reaction attack on the proposal of Wagner and Magyarik.* Let us assume that there is an element $0 \in \Sigma$ such that for all $\sigma, \tau \in \Sigma \setminus \{0\}$ the words $w_0 w_\sigma$ and $w_\tau w_0$ are inequivalent in G and $\widetilde{G}$. If $W(\Sigma)$ is not explicitly constructed to withstand this condition, then the existence of such an element w_0 seems plausible. Moreover, as for an alphabet Σ of moderate size the attacker can simply exhaust all possible choices of w_0, we consider w_0 as known to the attacker.

Also, we assume to have at hand a set of words $A \in (X \cup X^{-1})^*$ such that

- an exhaustive search over A is feasible, and
- from its subset $\bar{S} = \{a \in A \mid a \sim e \text{ in } \widetilde{G}\}$ one can derive a set $\bar{\bar{S}}$ so that $(X, \bar{\bar{S}})$ is a presentation of $\widetilde{G}$ (or either of another quotient of G that also provides a valid private key—see [**WM85**, Section 4.2, Attack (b)]).

Our goal is to find $\bar{S}$ by making use of a *validity check* oracle $\mathcal{O}$: given a word $w \in (X \cup X^{-1})^*$, we have $\mathcal{O}(w) = 1$ if w corresponds to a correct ciphertext, (i.e., if in $\widetilde{G}$ we have $w \sim w_\sigma$ for some $\sigma \in \Sigma$,) and we have $\mathcal{O}(w) = 0$ otherwise. To find $\bar{S}$ we perform an exhaustive search through A. Namely, for each $a \in A$ we send ≤ 2 queries to the oracle $\mathcal{O}$ (resp. a legitimate recipient) to decide whether $a \in \bar{S}$:

i.: aw_0:
- If $\mathcal{O}(aw_0) = 0$, then obviously $a \notin \bar{S}$.
- If $\mathcal{O}(aw_0) = 1$, then $a \in \bar{S}$, or in $\widetilde{G}$ we have $aw_0 \sim w_\sigma$ for some $\sigma \in \Sigma \setminus \{0\}$ (and hence $a \notin \bar{S}$). To distinguish these cases, a second query can be used:

ii.: $w_0 a$:
- If $\mathcal{O}(w_0 a) = 0$, then obviously $a \notin \bar{S}$.
- If $\mathcal{O}(w_0 a) = 1$, then $a \in \bar{S}$, or in $\widetilde{G}$ we have $w_0 a \sim w_\tau$ for some $\tau \in \Sigma \setminus \{0\}$ (and hence $a \notin \bar{S}$). In the latter case ($a \notin \bar{S}$) we conclude that $w_0 a w_0 \sim w_\tau w_0$. But from the previous query we know that the situation $a \notin \bar{S}$ occurs only if $aw_0 \sim w_\sigma$ for some $\sigma \in \Sigma \setminus \{0\}$, i.e., if $w_0 a w_0 \sim w_0 w_\sigma$ for some $\sigma \in \Sigma \setminus \{0\}$—in contradiction to $w_0 w_\sigma \not\sim$

$w_\tau w_0$ for $\sigma, \tau \in \Sigma \setminus \{0\}$. In summary, the situation $\mathcal{O}(w_0 a) = 1$ and $a \notin \bar{S}$ is impossible, and $\mathcal{O}(w_0 a) = 1$ implies $a \in \bar{S}$.

As explained in [**GVS01**], for the concrete example given by Wagner and Magyarik, this simple approach allows to retrieve the complete secret key. Thus, this attack demonstrates the proposed scheme as insecure in the sense of IND-VCA. Another, more recent, example of a proposal originating in combinatorial group theory and not taking into account "active" adversaries is due to Birget et al.

3.2. Making use of a decryption oracle. In [**BMS04, BMS05**] Birget et al. present a security analysis of the scheme of Wagner and Magyarik and introduce a new design inspired in [**WM85**] and using a similar problem as a base. While the designers pay some attention to VCA attacks, the encryption scheme may as well (if parameters are not chosen carefully enough) be vulnerable to very simple CPA attacks. In any case, it offers virtually no protection against a CCA2-adversary as proposed in [**GV05**]:

The encryption scheme can be briefly described as follows: Let $G = (X, R)$ be a finitely presented group, and fix a (faithful and transitive) action of G on the set $\{0, 1, 2\}^*$. We will denote by $(\hat{x})g$ the resulting word of the action of $g \in G$ on a word $\hat{x} \in \{0, 1, 2\}^*$.

Fix $x, z^0, z^1 \in \{0, 1, 2\}^*$. The public key consists of two finite sequences of small sets of words, $S_0 = [Z_1^0, \ldots, Z_m^0]$, $S_1 = [Z_1^1, \ldots, Z_r^1]$, with $Z_i^j \subseteq (X \cup X^{-1})^*$ such that words in $Z_1^0 \cdots Z_m^0$ (resp. $Z_1^1 \cdots Z_r^1$) represent group elements moving x to z^0 (z^1). The authors suggest carrying out the encryption of a bit $b \in \{0, 1\}$, simply constructing from S_b a group element g_b such that $(x)g_b = z^b$.

The private key is (x, z^0, z^1). Thus, for decrypting a given word $c \in (X \cup X^{-1})^*$, the receiver computes $(x)c$ and if $(x)c = z^b$ retrieves the corresponding bit b.

The "security" of this scheme is supposed to rely on the hardness of the following premise problem $\mathcal{P}$:
Given a a word w over $X \cup X^{-1}$ which is either equivalent to a word in $Z_1^1 \cdots Z_m^1$ or to a word in $Z_1^0 \cdots Z_m^0$, decide which of the two is actually the case.
Moreover, the authors provide in [**BMS04, BMS05**] several hints on how to select the parameters in order to guarantee the hardness of the problem.

Indeed, some steps in the above design should be carried out with special care in order to avoid simple VCA or even CPA attacks. In any case, in the present form the scheme can be easily proven to be insecure against CCA2-adversaries: Let $\mathcal{A}$ be an attacker having access to a decryption oracle $\mathcal{O}$ (with the usual limitation in the CCA2 setting), and denote by c a target ciphertext, $\mathcal{A}$ would like to decrypt. So $\mathcal{A}$ may feed $\mathcal{O}$ with any word in $(X \cup X^{-1})^* \setminus \{c\}$. In particular, $\mathcal{A}$ can simply choose two different words $w, \hat{w} \in Z_m^0$ and query the oracle $\mathcal{O}$ with the word $\hat{w} w^{-1} c$. Then $\mathcal{O}$ outputs a correct decryption b of c, as

$$x \xrightarrow{\hat{w}} z_1^0 \xrightarrow{w^{-1}} x \xrightarrow{c} z^b.$$

This attack is, again, independent of the hardness of the (claimed) underlying premise problem.

Note that one of the reasons why the above simple attacks succeed, is that an adversary can always construct new valid ciphertexts from given ones (and thus, VCA and CCA oracles are useful tools). To be fair one should also remark that the above proposals were made at a somewhat intuitive level, and did not specify the concrete security goals they were aiming at and which adversarial capabilities were taken into account. As the framework of provable security is geared to handle exactly this kind of questions, it seems natural to apply these techniques, too, when trying to base an encryption scheme on combinatorial group theory. As illustrated by the results in [**KLC**$^+$**00**], making use of the random oracle model, combining combinatorial group theory with provable security is possible. From a conceptual point of view, it is of course desirable to have a provably secure construction in the standard model. A natural starting point for achieving this, is a construction due to Cramer and Shoup. More recently, another efficient construction of a provably secure encryption scheme in the standard model has been proposed in [**CHK04**]. However, it seems harder to exploit this construction for the design of combinatorial group theoretical cryptographic tools.

4. Provable security for group based public key encryption schemes: the Cramer and Shoup construction

In [**CS98**], Cramer and Shoup presented a practical encryption scheme that could be proven IND-CCA secure in the standard model. Later, in [**CS01, CS02**] the authors generalized that construction introducing a methodology for deriving IND-CCA secure schemes based on a new mathematical tool for cryptography: universal projective hash families.

4.1. Universal projective hash families.
We recall (informally) the basic definitions given in [**CS02**]:

Let $H = \{H_k : X \longrightarrow \Pi\}_{k \in K}$ be a family of maps from a set X into a set Π (all sets considered are non-empty and finite), and let $\alpha : K \longrightarrow S$ be a map from K into another set S. The value $\alpha(k)$ can be seen as partial information about the map H_k, and is called the projection of H_k. A *projective hash family* (PHF) for a subset L of X is obtained whenever $\alpha(k)$ completely determines the restriction of H_k to L, i.e., for all $x \in L$ and $k_1, k_2 \in K$ the equality $\alpha(k_1) = \alpha(k_2)$ implies $H_{k_1}(x) = H_{k_2}(x)$. Typically, in order to compute the value $H_k(x)$ for an element $x \in L$ given $\alpha(k)$, one needs to be able to prove that x is actually in L; more precisely, $H_k(x)$ is usually derived from $\alpha(k), x$ and a *witness* for x, w, that acts as a proof of $x \in L$.

Depending on the amount of information on the behavior of H_k on $X \setminus L$ given by $\alpha(k)$, the PHF is called universal, smooth, 2-universal or strongly 2-universal. More precisely, let us consider that $k \in K$ is selected at random and suppose that $\alpha(k)$ is known. Let x be an arbitrary element in $X \setminus L$. The PHF is *universal* if the probability of correctly guessing $H_k(x)$ is negligible. The PHF is *2-universal* if this probability remains negligible even knowing $H_k(x^*)$ for a particular $x^* \neq x$. If the probability distribution of $H_k(x)$ is statistically close to the uniform distribution in K, the PHF is *smooth*. Finally, if this smoothness property holds even knowing $H_k(x^*)$ for $x^* \neq x$, then the PHF is *strongly 2-universal*.

4.2. The Cramer and Shoup encryption scheme.
In the public key encryption scheme proposed by Cramer and Shoup [**CS01, CS02**] a message $m \in \Pi$ is encrypted by using $H_k(x)$ for a random $x \in L$ as a one time pad, where H_k is a smooth PHF. The value of k is the secret key, $\alpha(k)$ is the public key and $(x, m \cdot H_k(x))$ is the encryption of m (implicitly, it is assumed that Π is a group

where elements can be efficiently inverted and multiplied). Clearly, the holder of k can retrieve $H_k(x)$, and therewith the message.

In order to enable the computation of ciphertexts, one requires the existence of an efficient algorithm, the *public evaluation algorithm*, computing the (unique) value $H_k(x)$ from $\alpha(k)$ and a *witness* w for $x \in L$.

Note that, under the assumption that random elements in L are hard to distinguish from random elements in $X \setminus L$, i.e., the subset membership problem is hard, there is no way for a polynomially bounded adversary to distinguish between a well-formed ciphertext and a fake ciphertext obtained by choosing $x \in X \setminus L$ instead of $x \in L$. In particular, given an element $x \in X$, even if the adversary knows $x \in L$, he will not be able to compute a witness for it and retrieve $H_k(x)$ from the public information. Moreover, due to the smoothness of the PHF, since k is unknown, $H_k(x)$ is close to be uniformly distributed on Π, so the message is nearly perfectly hidden. Therefore, no information about the plaintext can be obtained in polynomial time by a passive adversary.

IND-CCA security (i.e., security against active adversaries) is achieved by appending to the ciphertext a tag that acts as a proof of integrity. The tag is defined through an additional 2-universal PHF, and the new elements k' and $\alpha'(k')$ should be added to the public and secret keys. Roughly speaking, this tag prevents anyone ignoring m or a suitable witness for x from constructing valid ciphertexts; thus, a CCA oracle is of no help to the adversary, for he cannot construct anything "useful" to feed it with. More precisely the tag is defined as $t = H'_{k'}(c)$, where $c = (x, m \cdot H_k(x))$. As above, t is computed during encryption by means of the public evaluation algorithm with inputs $\alpha'(k')$, c and w. During decryption of (c, t), firstly t is checked to be equal to $H'_{k'}(c)$. If so, c is decrypted as above, otherwise the ciphertext is considered as corrupted and it is rejected.

4.3. Group action based projective hash families. In search of suitable projective hash families that could be used for their above mentioned scheme, Cramer and Shoup designed a construction method that took as a starting point certain atomic abelian group theoretic tools, defined as *group systems* [**CS01, CS02**]. In [**GVMSV05, GVV05**] the theory of Cramer and Shoup has been reproduced from a more general starting point, allowing for the use of a wider range of algebraic tools. Although being focused on finite groups, we think this approach to be of interest when aiming at constructions basing on combinatorial group theory: Already without the additional obstructions caused by the infinity of the underlying element set, building a provably secure construction on a non-abelian group turns out to be a non-trivial task.

The new construction works as follows. Let X be a finite set and consider a finite (not necessarily abelian) group H left-acting on X. Thus, each element $\phi \in H$ can be seen as an element of the symmetric group on X, and for all $\phi_1, \phi_2 \in H$ and $x \in X$, $(\phi_1 \phi_2)x = \phi_1(\phi_2 x)$. Moreover, denote by S some finite group and by $\alpha : H \longrightarrow S$ a group homomorphism. Note that for any $\phi \in H$, $\alpha(\phi)$ gives some (limited) information about ϕ, and thus α provides some partial information about the action of H on X. Hence, α plays the same role as in a PHF. Such X, H, S and α form a *group action system*.

Let us consider the action of $\ker\alpha$ on X, defined by the action of H. For any $x \in X$, let us denote by $[x]$ the $(\ker\alpha)$-orbit of x, i.e.,

$$[x] := \{\phi x \mid \phi \in \ker\alpha\}.$$

Note that the action of H on the set of points that remain fixed by $\ker\alpha$ is completely determined by α. Let us thus define

$$L := X^{\ker\alpha} = \{x \in X \mid |[x]| = 1\},$$

that is $L = \{x \in X \mid [x] = \{x\}\}$, and observe that $\ker\alpha \subseteq \mathrm{Stab}_H(L)$ although they are not necessarily equal. Also, H leaves L invariant, as for any $\phi \in H$ and $x \in L$, ϕx is fixed by all $\psi \in \ker\alpha$, as there exists $\rho \in \ker\alpha$ s.t. $\psi\phi = \phi\rho$ and thus $\psi\phi x = \phi\rho x = \phi x$.

As in [**CS01, CS02, GVMSV05**], we outline the construction of a projective hash family from a group action system. To this aim, we let $K = H$ and $\Pi = X$. Noting that $\alpha(\phi)$ determines the action of ϕ on L completely, it is easy to see that H and α define a PHF for $L \subset X$. We will refer to this PHF as a *group action projective hash family* (AcPHF). As our aim is to construct cryptographically useful PHFs, the system above will be useful for us if α gives little information about the action of H on $X \setminus L$; thus, we will be particularly interested in those systems for which the $(\ker\alpha)$-orbits of elements in $X \setminus L$ are large. We say that the group action system is *p-diverse*, for an integer $p > 1$, if $|[x]| \geq p$ for all $x \in X \setminus L$. It is proven in [**GVV05**] that if a group action system is p-diverse for a large enough p, then the corresponding AcPHF is universal. Moreover, for such a universal projective hash family there is a dedicated upgrading method that allows for the efficient construction of a 2-universal PHF.

Also, finding p-diverse group action systems can for instance be done by the following observation: p-diversity is guaranteed for p being the smallest prime dividing $|\ker\alpha|$. Other construction methods can be exploited for concrete settings (see [**GVMSV05, GVV05**]). Under the assumption that the underlying subset membership problem $L \subset X$ is hard, the AcPHF can be used to build an IND-CCA secure encryption scheme following Cramer and Shoup's construction. Note that for this purpose a public evaluation algorithm for $\phi(x)$ from $\alpha(\phi)$, $x \in L$ and a witness w should exist.

4.3.1. *An abelian example.* Cramer and Shoup presented in [**CS01, CS02**] group theoretic constructions of universal projective hash families based on abelian groups. They define an *(abelian) group system* from the abelian groups X, L and Π, where L is a proper subgroup of X, and a subgroup $\mathcal{H}$ of the homomorphism group between X and Π. Abelian group systems also yield natural examples of AcPHFs, based on the observation that $\mathrm{Hom}(X, \Pi)$ left-acts on $X \times \Pi$ in the natural way: $h(x, \pi) = (x, h(x)\pi)$. Thus, all concrete constructions from [**CS01, CS02**] yield examples in our new setting.

4.3.2. *An example using linear groups.* Let us consider the vector space $X = K^n$, where K is the finite field with q elements, and $\{\alpha_1, \ldots, \alpha_n\}$ a K-basis of X. Clearly, $GL(n,q)$ left-acts on X. Consider H a subgroup of $GL(n,q)$, leaving a d-dimensional subspace L of X invariant. For a given $M \in GL(n,q)$, denote by M_d the matrix representing the linear transformation induced by M on L in a certain basis $\{\beta_1, \ldots, \beta_d\}$. Clearly,

$$\alpha: H \longrightarrow GL(d,q)$$
$$M \longmapsto M_d,$$

is a group homomorphism.

Now, diversity of the group action system can be proven if H is chosen with care (ideally, $|H|$ or better $|\ker \alpha|$ should only have large prime divisors). However, depending on the information about X and H made public, it would be difficult to hide the subspace L. Hence, we have to encode the elements $x \in X$ and $M \in H$ in a way that the computation of $M \cdot x$ is infeasible from the encodings of x and M, but it is still possible if at least one of x and M are known in the clear.

Unfortunately, so far no new practically convincing public key encryption scheme could be derived from an AcPHF building on a non-abelian group. However, the already achieved results towards a provably secure encryption scheme in the standard model building on non-abelian groups certainly justify further research along this line of thought.

5. Conclusion

The above discussion reiterates the point that for exploring the cryptographic potential of combinatorial group theory, it is essential to take established cryptographic security models and tools into account. Besides helping to avoid misconceptions about underlying assumptions and security goals, the framework of provable security can help to provoke "mathematically meaningful" attacks, i.e., attacks indeed aiming at the underlying theoretical questions. As a possible starting point for future research towards provable security for cryptographic schemes arising from combinatorial group theory, we outlined a non-abelian construction basing on a well-known (abelian) design methodology due to Cramer and Shoup.

References

[AAG99] Iris Anshel, Michael Anshel, and Dorian Goldfeld, *An Algebraic Method for Public-Key Cryptography*, Mathematical Research Letters **6** (1999), 287–291.

[ATS03] P. Jeyanthi Abisha, D. Gnanaraj Thomas, and K. G. Subramanian, *Public Key Cryptosystems Based on Free Partially Commutative Monoids and Groups*, Proceedings of INDOCRYPT 2003 (Thomas Johansson and Subhamoy Maitra, eds.), Lecture Notes in Computer Science, vol. 2904, Springer, 2003, pp. 218–227.

[BDPR98] Mihir Bellare, Anand Desai, David Pointcheval, and Philip Rogaway, *Relations Among Notions of Security for Public-Key Encryption Schemes*, Advances in Cryptology — CRYPTO '98 (Hugo Krawczyk, ed.), vol. 1462, Springer, 1998, pp. 26–46.

[BMS04] Jean-Camille Birget, Spyros S. Magliveras, and Michal Sramka, *On public-key cryptosystems based on combinatorial group theory*, Presented at the 4th Central European Conference on Cryptology, *WARTACRYPT 2004*, July 2004.

[BMS05] _____, *On public-key cryptosystems based on combinatorial group theory*, Cryptology ePrint Archive: Report 2005/070, 2005, available at http://eprint.iacr.org/2005/070/.

[CHK04] Ran Canetti, Shai Halevi, and Jonathan Katz, *Chosen-Ciphertext Security from Identity-Based Encryption*, Advances in Cryptology — EUROCRYPT 2004 (Christian Cachin and Jan Camenisch, eds.), Lecture Notes in Computer Science, vol. 3027, Springer, 2004, pp. 207–222.

[CKN03] Ran Canetti, Hugo Krawczyk, and Jesper Buus Nielsen, *Relaxing Chosen-Ciphertext Security*, Advances in Cryptology - CRYPTO 2003 (Dan Boneh, ed.), Lecture Notes in Computer Science, vol. 2729, Springer, 2003, pp. 565–582.

[CS98] Ronald Cramer and Victor Shoup, *A Practical Public Key Cryptosystem Provably Secure against Adaptive Chosen Ciphertext Attack*, Proceedings of CRYPTO'98 (H. Krawczyk, ed.), LNCS, vol. 1462, Springer, 1998, pp. 13–25.

[CS01] _____, *Universal Hash Proofs and a Paradigm for Adaptive Chosen Ciphertext Secure Public-Key Encryption*, Cryptology ePrint Archive: Report 2001/085, 2001, available at http://eprint.iacr.org/2001/085/.

[CS02] _____, *Universal Hash Proofs and a Paradigm for Adaptive Chosen Ciphertext Secure Public-Key Encryption*, Advances in Cryptology — EUROCRYPT 2002 (Lars Knudsen, ed.), Lecture Notes in Computer Science, vol. 2332, Springer, 2002, pp. 45–64.

[DDN00] Danny Dolev, Cynthia Dwork, and Moni Naor, *Non-malleable cryptography*, SIAM Journal on Computing **30** (2000), 391–437.

[Deh04] Patrick Dehornoy, *Braid-based cryptography*, Contemporary Mathematics **30** (2004), 5–33.

[dVP04] Françoise Levy dit Vehel and Ludovic Perret, *Attacks on Public Key Cryptosystems Based on Free Partially Commutative Monoids and Groups*, Progress in Cryptology - INDOCRYPT 2004: 5th International Conference on Cryptology in India (Anne Canteaut and Kapaleeswaran Viswanathan, eds.), Lecture Notes in Computer Science, vol. 3348, Springer, 2004, pp. 275–289.

[GM84] Shafi Goldwasser and Silvio Micali, *Probabilistic Encryption*, Journal of Computer and System Sciences **28** (1984), no. 2, 270–299.

[GV05] María Isabel González Vasco, *On the security of a group based public key cryptosystem*, Proceedings of *Workshop on Mathematical Problems and Techniques in Cryptology* (Bellatera, Spain), Centre de Recerca Matemàtica, 2005.

[GVHMS04] María Isabel González Vasco, Dennis Hofheinz, Consuelo Martínez, and Rainer Steinwandt, *On the security of two public key cryptosystems using non-abelian groups*, Designs, Codes and Cryptography (Special Issue: Proceedings of the Third Pythagorean Conference) **32** (2004), 207–216.

[GVMSV05] María Isabel González Vasco, Consuelo Martínez, Rainer Steinwandt, and Jorge L. Villar, *A new Cramer-Shoup like methodology for group based provably secure schemes*, Proceedings of the 2nd Theory of Cryptography Conference TCC 2005 (Joe Kilian, ed.), Lecture Notes in Computer Science, vol. 3378, Springer, 2005, pp. 495–509.

[GVS01] María Isabel González Vasco and Rainer Steinwandt, *Clouds over a Public Key Cryptosystem Based on Lyndon Words*, Information Processing Letters **80** (2001), 239–242.

[GVS04] _____, *A Reaction Attack on a Public Key Cryptosystem Based on the Word Problem*, Applicable Algebra in Engineering, Communication and Computing **14** (2004), no. 5, 335–340.

[GVSar] _____, *Pitfalls in public key cryptosystems based on free partially commutative monoids and groups*, Applied Mathematics Letters (to appear).

[GVV05] María Isabel González Vasco and Jorge L. Villar, *In search of mathematical primitives for deriving universal projective hash families*, Tech. report, Centre de Recerca Matemàtica, July 2005, available at http://www.crm.es/Publications/05/pr641.pdf.

[GZ91] Max Garzon and Yechezkel Zalcstein, *The Complexity of Grigorchuk groups with application to cryptography*, Theoretical Computer Science **88** (1991), 83–98.

[HGS99] Chris Hall, Ian Goldberg, and Bruce Schneider, *Reaction Attacks Against Several Public-Key Cryptosystems*, Information and Communication Security, Second International Conference, ICICS'99 (Vijay Varadharajan and Yi Mu, eds.), Lecture Notes in Computer Science, vol. 1726, Springer, 1999, pp. 2–12.

[HMQSar] Dennis Hofheinz, Jörn Müller-Quade, and Rainer Steinwandt, *On modeling IND-CCA security in cryptographic protocols*, Tatra Mountains Mathematical Publications (to appear).

[KLC$^+$00] Ki Hyoung Ko, Sang Jin Lee, Jung Hee Cheon, Jae Woo Han, Ju sung Kang, and Choonsik Park, *New Public-Key Cryptosystem Using Braid Groups*, Advances in Cryptology — CRYPTO 2000 (Mihir Bellare, ed.), Lecture Notes in Computer Science, vol. 1880, Springer, 2000, pp. 166–183.

[KMar]   Neal Koblitz and Alfred Menezes, *Another Look at Provable Security*, Journal of Cryptology (to appear).

[Per05]   Ludovic Perret, *A Chosen Ciphertext Attack on a Public Key Cryptosystem Based on Lyndon Words*, March 2005, see also Cryptology ePrint Archive, http://eprint.iacr.org/2005/014.

[Pet03]   George Petrides, *Cryptanalysis of the Public Key Cryptosystem Based on the Word Problem on the Grigorchuk Groups*, Cryptography and Coding (Kenneth G. Paterson, ed.), Lecture Notes in Computer Science, vol. 2898, Springer, 2003, pp. 234–244.

[Poi04]   David Pointcheval, *Advanced Course on Contemporary Cryptology*, ch. Provable Security for Public Key Schemes, Centre de Recerca Matemàtica, February 2004, available at http://di.ens.fr/~pointche/pub.php.

[Sho98]   Victor Shoup, *Why Chosen Ciphertext Security Matters*, Tech. Report RZ 3076 (#93122), IBM Research Division, Zurich Research Laboratory, November 1998, available at http://www.shoup.net/papers/expo.pdf.

[SM90]   Rani Siromoney and Lisa Mathew, *A Public Key Cryptosystem Based on Lyndon Words*, Information Processing Letters **35** (1990), 33–36.

[STAS02] S.C. Samuel, D. Gnanaraj Thomas, P. Jeyanthi Abisha, and K.G. Subramanian, *Tree Replacement and Public Key Cryptosystem*, Proceedings of INDOCRYPT 2002 (Alfred Menezes and Palash Sarkar, eds.), Lecture Notes in Computer Science, vol. 2551, Springer, 2002, pp. 71–78.

[Ste03]   Jacques Stern, *Why Provable Security Matters?*, Advances in Cryptology: Proceedings of EUROCRYPT 2003 (Eli Biham, ed.), Lecture Notes in Computer Science, vol. 2656, Springer, 2003, pp. 449–461.

[WM85]   Neal R. Wagner and Marianne R. Magyarik, *A Public Key Cryptosystem Based on the Word Problem*, Advances in Cryptology—CRYPTO '84 (G. R. Blakley and David Chaum, eds.), Lecture Notes in Computer Science, vol. 196, Springer, 1985, pp. 19–36.

ÁREA DE MATEMÁTICA APLICADA, UNIVERSIDAD REY JUAN CARLOS, C/ TULIPÁN S/N, 28933 MÓSTOLES, MADRID, SPAIN
E-mail address: mariaisabel.vasco@urjc.es

DEPARTMENT OF MATHEMATICAL SCIENCES, FLORIDA ATLANTIC UNIVERSITY, 777 GLADES ROAD, BOCA RATON, FL 33431, USA
E-mail address: rsteinwa@fau.edu

DEPARTAMENTO DE MATEMÁTICA APLICADA IV, UNIVERSITAT POLITÉCNICA DE CATALUNYA, CAMPUS NORD, C/JORDI GIRONA, 1–3, 08034 BARCELONA, SPAIN
E-mail address: jvillar@ma4.upc.edu

Constructions in public-key cryptography over matrix groups

Dima Grigoriev and Ilia Ponomarenko

ABSTRACT. The purpose of the paper is to give new key agreement protocols (a multi-party extension of the protocol due to Anshel-Anshel-Goldfeld and a generalization of the Diffie-Hellman protocol from abelian to solvable groups). They as well as a number of homomorphic public-key cryptosystems rely on difficulty of the conjugacy and membership problems for subgroups of a given group. To support all of them we present a general technique to produce a family of instances being matrix groups (over finite commutative rings) which play a role for these schemes similar to the groups Z_n^* in the existing cryptographic constructions like RSA or discrete logarithm.

Introduction

One of the oldest cryptographical problems consists in constructing of a key agreement protocol. Roughly speaking it is a multi-party algorithm, defined by a sequence of steps, specifying the actions of two or more parties in order a shared secret becomes available to two or more parties. Probably the first such procedure based on abelian groups is due to Diffie-Hellman (see [**GB01**]). In fact, it concerns automorphisms of abelian (even cyclic) groups induced by taking to a power. Some generalizations of this protocol to non-abelian groups (in particular, the matrix groups over some rings) were suggested in [**PKHK**] where security was based on an analog of the discrete logarithm problems in groups of inner automorphisms. Certain variations of the Diffie-Hellman systems over the braid groups were described in [**KLCHKP**]; there several trapdoor one-way functions connected with the conjugacy and the taking root problems in the braid groups were proposed. A general scheme for constructing key agreement protocols based on algebraic structures was proposed in [**AAG99**]. In principle, it enables one to construct such protocols for non-abelian groups and their automorphisms induced by conjugations. In this paper we generalize to the non-abelian case the Diffie-Hellman protocol, construct

---

1991 *Mathematics Subject Classification.* Primary 68Q17; Secondary 20C40, 20H25.

Key words and phrases. Public-key cryptosystems, matrix groups.

The second author was supported by RFFI, grants NSH-4329.2006.1, #05-01-00899. The paper was done during the stay of the author at the Mathematical Institute of the University of Rennes.

multi party procedure for the protocol [**AAG99**], and analyze the security of both protocols realized in matrix groups over rings.

The question on finding probabilistic public-key cryptosystems in which the decryption function has a homomorphic property goes back to [**RAD78**] (see also [**FM91**]). In such a cryptosystem the spaces of messages and of ciphertexts are algebraic structures G and H and the decryption function $D : G \to H$ is a homomorphism. A number of such cryptosystems is known for abelian groups, e.g. the quadratic residue cryptosystem [**GB01**] and its generalization for highest residues [**NS98**] (see also an overview in [**GP**]). In most of them the security is based on the intractability of number-theoretical problems close to the integer factoring. Recently, several homomorphic cryptosystems were constructed for infinite (but finitely presented) groups, see [**GP, GP04**] and references there.

The third problem considered in this paper is how to produce instances for cryptosystems based on computations with matrix groups over rings. In contrast to numerous theoretical cryptosystems where there is a lot of efficient algorithms to generate integers with given properties (e.g., the pairs of two distinct large primes of the same bit size used in the quadratic residue cryptosystem), it is not clear a priory how to find efficiently matrix groups in which some problems (like membership or conjugacy) arising in cryptography are computationally difficult. We propose a general scheme for solving this problem and give a specialization of this scheme for matrix groups over finite commutative rings.

In Section 1 we study key agreement protocols between two parties (named usually Alice and Bob) and their extensions to several parties. The security of the Diffie-Hellman protocol relies on the difficulty of the following *transporter problem*: having an action $G \times V \to V$ of a group G on a set V for given $u, v \in V$ to find $g \in G$ (provided that it does exist) such that $(g, u) \mapsto v$. In case of V being a cyclic group of order n and G being a group acting on V by taking a power, one arrives to the discrete logarithm problem (usually, n is taken to be prime). The security of the key agreement protocol of [**AAG99**] (see also Subsection 1.1) relies on the difficulty of the conjugacy problem with respect to a subgroup of G. In Subsection 1.1 we extend the construction of [**AAG99**] to *multi-party* key agreement protocol. Then in Subsection 1.2 we design another generalization of the Diffie-Hellman protocol to actions of groups G which satisfy a certain *identity*. Clearly, any abelian group satisfies the identity $aba^{-1}b^{-1} = 1$ and more generally, any solvable group with a fixed length of its derived series satisfies an appropriate commutator identity. The security of our protocol again relies on the difficulty of the transporter problem for a suitable action of G.

In Section 2 we consider homomorphic public-key cryptosystems (see e.g. [**GP**]) in which the decrypting function (known to Alice) is a group homomorphism $f : G \to H$ where the groups H, G play the roles of the spaces of plain and ciphertext messages, respectively. Usually, the security of a homomorphic cryptosystem relies on the difficulty of the problem of the membership to a normal subgroup of G (here, the kernel of f). For example, in a homomorphic cryptosystem from [**GP04**] G was a subgroup of the modular group $SL_2(\mathbb{Z})$ and the security of this cryptosystem relied on the difficulty of a certain membership problem to a subgroup of the modular group.

The crucial role in the classical cryptographic constructions (like RSA, discrete logarithm or quadratic residue [**GB01**]) plays the natural action of the group

Aut($\mathbb{Z}_n^*$) on the group $\mathbb{Z}_n^*$. So, varying n one gets a mass pool of instances for cryptographic primitives. This action is a special case of the natural action of the group $Aut_R(V)$ (viewed as a matrix group) on the free module V over the ring R. In this paper we propose a construction of a pool of matrix groups instances for cryptographic primitives (Subsection 3.2). The security of these instances relies on the difficulty of certain problems on matrix groups (e.g. the membership to a subgroup or the conjugacy with respect to a subgroup). For the complexity of such problems few results were established in case of matrix groups over fields [**BB93, KS03**]; for matrix groups over arbitrary rings much less is known. We note also that matrix groups were mentioned in [**EK**] as candidates for groups with a difficult conjugacy problem.

The common way in cryptography of producing a trapdoor and a cryptosystem, is to generate a private key departing from a pair of primes p, q, while their product $n = pq$ plays the role of a public key. In our scheme (see Subsection 3.1) as a private key we take a rooted tree (called a derivation tree) whose leaves being furnished with specially chosen (non-abelian, in general) groups. We assume that Alice has in possession such representations of these groups which allow her to solve efficiently a problem lying in the background of a cryptosystem (like membership or conjugacy). Internal vertices of the tree are endowed with certain operations on groups which allow one to assign recursively a group to each vertex of the tree starting with its leaves. At the end of the recursion a group is assigned to the root, and this group plays the role of a public key. This scheme is also modified to produce a "secret" homomorphism of matrix groups with a private key being a derivation tree for this homomorphism (similar to the derivation tree of the group). In Subsection 3.2 we give a realization of this general scheme in finite matrix groups.

The similarity of the common constructions in cryptography based on commutative groups (say, $\mathbb{Z}_n^*$) with our construction (relying on finite matrix groups) allows us to call the latter type of constructions the *non-commutative cryptography*.

1. Group-theoretical key agreement protocol

1.1. A multi-party protocol. The following group-theoretical variant of key agreement two-party protocol was proposed in [**AAG99**]. Let G be a group, and to two parties A and B are assigned their subgroups

(1) $$G_A = \langle a_1, \ldots, a_m \rangle, \quad G_B = \langle b_1, \ldots, b_n \rangle.$$

The group G and the elements a_i, b_j are publically known. The parties A and B choose secret elements $a \in G_A$ and $b \in G_B$ and transmit to each other the collections

$$X_B = \{a^{-1}b_j a\}_{j=1}^n, \quad X_A = \{b^{-1}a_i b\}_{i=1}^m,$$

respectively. Since A (resp. B) has a representation of the element a (resp. b) via generators $a_1, \ldots, a_m$ (resp. $b_1, \ldots, b_n$), then A (resp. B) can compute a representation of the element $b^{-1}ab$ (resp. $a^{-1}ba$) via elements of the set X_A (resp. X_B). Thus, A and B have a common key

$$a^{-1}(b^{-1}ab) = [a, b] = (a^{-1}ba)^{-1}b.$$

An obvious necessary condition for this protocol to be secure is that the set of all such commutators with $a \in G_A$, and $b \in G_B$ should contain at least two elements.

Let us describe a generalization of the group-theoretical key agreement protocol for s parties with $s \geq 2$ and a single public communicating channel. In contrast to the straightforward algorithm having a quadratic complexity, we give an algorithm the complexity of which is linear in s. Without loss of generality we assume that $s = 2^t$ for some $t \geq 1$, for otherwise in the recursive construction below we divide the parties into two unequal subsets which leads just to a slight changing the notation. As in the case $s = 2$ the groups $G_1, \ldots, G_s \subset G$ of the parties are given publically by their sets of generators. At the initial step the ith party chooses a secret key $a_i \in G_i$, $i = 1, \ldots, s$. Let S_1 and S_2 be disjoint $s/2$-subsets of the set $\{1, \ldots, s\}$. Then given $u = 1, 2$ the parties from S_u recursively construct the common key $K_u \in G$, such that for all $i \in S_u$ there exist integers $\varepsilon_{i,j} \in \{-1, +1\}$ and $1 \leq m_i \leq s/2$, $1 \leq j \leq m_i$, and certain elements $B_{i,1}, \ldots, B_{i,m_i} \in \langle \{a_l : l \in S_{u,i}\} \rangle$ with $S_{u,i} = S_u \setminus \{i\}$, for which we have

$$K_u = (B_{i,1}^{-1} a_i^{\varepsilon_{i,1}} B_{i,1}) \cdots (B_{i,m_i}^{-1} a_i^{\varepsilon_{i,m_i}} B_{i,m_i}).$$

By recursion we can assume that the ith party knows the elements $B_{i,j}^{-1} a B_{i,j}$ for all j and for all chosen generators a of the group G_i (and thereby, it knows $B_{i,j}^{-1} a_i B_{i,j}$), but does not necessary know $B_{i,j}$. At this point the party $i \in S_u$ sends the elements $B_{i,j}^{-1} a B_{i,j}$ for all the chosen generators a of the group G_i to a certain party from the set $S_{u'}$ with $u' = 3 - u$ and asks for the elements $K_{u'}^{-1} B_{i,j}^{-1} a B_{i,j} K_{u'}$. Then for $u = 1$ the ith party computes the element

$$[K_1, K_2] = K_1^{-1}(K_2^{-1} K_1 K_2) =$$
$$K_1^{-1}(K_2^{-1}(B_{i,1}^{-1} a_i^{\varepsilon_{i,1}} B_{i,1}) K_2) \cdots (K_2^{-1}(B_{i,m_i}^{-1} a_i^{\varepsilon_{i,m_i}} B_{i,m_i}) K_2).$$

Similarly, for $u = 2$ the ith party computes the element $[K_1, K_2] = (K_1^{-1} K_2 K_1)^{-1} K_2$. Thus, this element can be chosen as the common key for all parties. It is easy to see that the ith party computes the common key in $O(s|a_i|)$ operations in the group G, where $|a_i|$ denotes the length of the word a_i in the chosen generators of the group G_i.

1.2. A new protocol. In this subsection we define a new group-theoretical two party key agreement protocol that can be viewed as a non-commutative generalization of the Diffie-Hellman protocol (see [**GB01**]).

Let G be a group acting on a set X so that given $(x, g) \in X \times G$ the image x^g of x with respect to g can be efficiently computed. Two parties A and B going to choose a secret common key from X, fix publically subgroups G_A, G_B of the group G and two words

$$W_A(u_A, u_B) = u_A^{a_{1,1}} u_B^{b_{1,1}} \cdots u_A^{a_{1,m_1}}, \qquad W_B(u_A, u_B) = u_B^{b_{2,1}} u_A^{a_{2,1}} \cdots u_B^{b_{2,m_2}}$$

of the free group F_2 with two free generators u_A, u_B such that

(W1) $m_1, m_2 \in \mathbb{N}$, $a_{i,j}, b_{i,j} \in \mathbb{Z}$ for all i, j, and $a_{1,m_1} \neq 0$, $b_{2,m_2} \neq 0$,

(W2) $W_A(g_A, g_B) = W_B(g_A, g_B)$ for all $(g_A, g_B) \in G_A \times G_B$.

The protocol begins with the choice of a publically known element $x_0 \in X$ and the secret elements $g_A \in G_A$ by the party A and $g_B \in G_B$ by the party B. Then during the communications the party A performs the following:

- At step 0 set $K_A = x_0$.
- At steps $i = 1, \ldots, m_1 - 1$ send $K_A^{g_A^{a_{1,i}}}$ and receive $K_A := K_A^{g_A^{a_{1,i}} g_B^{b_{1,i}}}$.
- At step $i = m_1$ set $K_A := K_A^{g_A^{a_{1,m_1}}}$.

The communications of the party B are defined similarly. Thus, at the end of the communication process due to condition (W2) the parties A and B have the common key
$$K_A = x_0^{W_A(g_A, g_B)} = x_0^{W_B(g_A, g_B)} = K_B.$$

For $X = \mathbb{Z}_p^*$ with p being a prime, $G = G_A = G_B$ being the group $\mathbb{Z}_{p-1}^* \cong \mathrm{Aut}(\mathbb{Z}_p^*)$ and $W_A(u_A, u_B) = u_B u_A$, $W_B(u_A, u_B) = u_A u_B$ we come to the Diffie-Hellman protocol.

This scheme can be easily realized for a solvable group G with bounded length n of the derived series of G. For example, one can take $G_A = G_B = G$ and choose the words $W_A = W_{A,n}$ and $W_B = W_{B,n}$ by induction on n as follows. If $n = 1$, then the group G is abelian and so conditions (W1) and (W2) are satisfied for
$$W_{A,1}(u_A, u_B) = u_B u_A, \qquad W_{B,1}(u_A, u_B) = u_A u_B.$$

For $n \geq 2$ the commutator $[g, h] = g^{-1} h^{-1} g h$ with arbitrary $g, h \in G$ belongs to the derived subgroup $G' = [G, G]$ of G (the derived length of G' equals $n - 1$). Assume by induction that conditions (W1) and (W2) are satisfied for the words $W_{A,n-1}$ and $W_{B,n-1}$. Then a straightforward checking shows that these conditions are also satisfied, for example, for the words
$$W_{A,n} = W_{A,n-1}([u_B, u_A], [u_A^{-1}, u_B^{-1}]), \qquad W_{B,n} = W_{B,n-1}([u_B, u_A], [u_A^{-1}, u_B^{-1}]).$$

Indeed, property (W2) is obvious. Next, one can verify by induction on $n \geq 1$ that the length (the number of letters) of the word $W_{A,n}$ (as well as $W_{B,n}$) equals $2 \cdot 4^{n-1}$. This means that there are no reductions in these words which implies property (W1).

More generally, to define $W_{A,n}$ and $W_{B,n}$ one can choose arbitrary words $W_1, W_2, W_3, W_4 \in W_Y$ where $Y = \{u_A, u_B\}$ and W_Y is the set of all words in the alphabet $Y^\pm$, and use $[W_1, W_2]$ and $[W_3, W_4]$ instead of $[u_A, u_B]$ and $[u_B^{-1}, u_A^{-1}]$, respectively. Certainly, to provide condition (1) one should guarantee that the words $W_{A,n-1}(u_A, u_B)$ (resp. $W_{B,n-1}(u_A, u_B)$) and W_2 (resp. W_4) must be terminated to u_A (resp. u_B). To avoid triviality we also should take $W_1, \ldots, W_4$ so that $W_{A,n}$ and $W_{B,n}$ would be nonidentity elements in the underlying free group.

Clearly, any realization of the above protocol is based on identities of the group G. In addition to commutator identities for solvable groups (see above) one can also use the identity $x^m = 1$ (that holds in any finite group the order of which is a divisor of m, and in the Burnside groups). In this case we can choose as the words W_A and W_B the prefix and the inverse of the suffix of the word $(u_A u_B)^m$, respectively, so that the prefix is terminated to u_A. In fact, as it was proved by B.Neumann any variety of groups can be given by a collection of identities such that the first of them is of the form $x^m = 1$ with m being a nonnegative integer, whereas the other ones are the elements of the commutant of the underlying free group (see [**MKS66**]).

We complete the subsection by making two remarks on the above protocol. First, the set X must be of superpolynomial size, for otherwise the key agreement scheme can be broken in polynomial time by the known permutation group theory technique (see [**L93**]). Second, the words W_A and W_B must be chosen so that the number of elements $W_A(g_A, g_B) = W_B(g_A, g_B)$ with $g_A, g_B \in G$ would be at least two.

1.3. On the security of the protocols.
In the above protocols we assume that all groups are given explicitly, e.g. by sets of generators, so that the group operations can be performed efficiently. Then the security of the first protocol is based on the intractability of the following problem (see [**SU06**]).

Subgroup Conjugation Search Problem (SCSP). *Given a group G, subgroups H_1, H_2 of G, and two elements $f, g \in H_1$, find an element $h \in H_2$ such that $f = h^{-1}gh$, provided that at least one such h exists.*

As usually in the cryptography, an efficient algorithm solving SCSP would break the protocol (but to break the protocol it is not necessary to solve SCSP). Such an algorithm does exist for $G = \mathrm{GL}(n, \mathbb{F}_q)$ where n is a natural number, $\mathbb{F}_q$ is a finite field of the order q, and the subalgebra $A(H_2)$ of the full matrix algebra $\mathrm{Mat}_n(\mathbb{F}_q)$ generated by the group H_2 is such that

$$A(H_2) \cap G = H_2.$$

Then for arbitrary H_1 the problem SCSP can be solved in probabilistic polynomial time (in n and in $\log q$) by the linear algebra technique, provided that n is less than $q/2$. Indeed, in this case the solution of the linear system $hf - gh = 0$ with respect to $h \in A(H_2)$ is an element of H_2 with a great probability. (From [**CY97**] it follows that in this case the problem SCSP can be solved efficiently even by a deterministic algorithm.)

It seems that the problem SCSP remains difficult when G is restricted to subgroups of the group $\mathrm{GL}(V, R)$ of all invertible R-linear transformations of the free R-module V where R is a finite commutative ring. To see this we consider the Linear Transporter Problem on the intractability of which the second protocol is based.

Linear Transporter Problem (LTP). *Let R be a commutative ring, V be an R-module and $G \leq \mathrm{GL}(V, R)$. Given $u \in V$ and $v \in u^G = \{u^g : g \in G\}$ find $g \in G$ such that $v = u^g$.*

A special case of (LTP) is the Discrete Logarithm Problem. Indeed, take $V = \mathbb{Z}_p^*$ with p being a prime. Then V can be considered as an one-dimensional module over the ring $R = \mathrm{End}(V) \cong \mathbb{Z}_{p-1}$ (with respect to taking the power $v \mapsto v^n$ where $v \in V$, $n \in \mathbb{Z}_{p-1}$). Choosing u to be a generator of the group V we come to the Discrete Logarithm Problem.

Preserving the notation of LTP set $T(V) = \{T_v : x \mapsto x + v,\ v, x \in V\}$ to be the translation group of the R-module V. Then obviously

$$v = u^g \Leftrightarrow T_v = g^{-1}T_u g, \qquad u, v \in V, \quad g \in \mathrm{GL}(V, R).$$

So, the problem LTP is the special case of the problem SCSP with $G = \mathrm{AGL}(V, R)$, $H_1 = T(V)$ and $H_2 = \mathrm{GL}(V, R)$. (Here $\mathrm{AGL}(V, R) = T(V)\mathrm{GL}(V, R)$ is the group of all affine transformations of V.) This shows that SCSP is at least as hard as LTP. In particular, this construction gives us a family of groups for which the problem SCSP turns to be at least as hard as the Discrete Logarithm Problem. A general technique to construct groups of this kind will be given in Section 3.

2. Homomorphic cryptosystems over groups

A homomorphic cryptosystem is a probabilistic public-key scheme (in the sense of [**GB01**]) in which the spaces of plaintext messages and ciphertexts are groups

H_k and G_k, respectively, depending on a security parameter k and such that its decryption function

(2) $$f_k : G_k \to H_k$$

is an epimorphism for all k. Usually, in a homomorphic cryptosystem the public key includes generator sets X_k and Y_k of the groups G_k and H_k, and some set $R_k \subset X_k$ such that $Y_k \subset f_k(R_k) = \{f_k(g) : g \in R_k\}$. Besides, it is assumed that there are publically known $k^{O(1)}$-time algorithms to solve the following problems:

- (H1) given two elements a, b of G_k (resp. H_k) find the element ab^{-1},
- (H2) given $y \in Y_k$ find an element of the set $R_k \cap f_k^{-1}(y)$,
- (H3) generate a random element of the group $\ker(f_k)$

where sizes of all elements are assumed to be at most k. Under these assumptions the encryption can be performed in time $k^{O(1)}$ as follows. First, given a message $h = y_1 \cdots y_m \in H_k$ with $y_i \in Y_k$ and m being a natural number at most $k^{O(1)}$, Bob computes in time polynomial in k an element $r = r_1 \cdots r_m \in G_k$ such that $r_i \in R_k$ and $f_k(r_i) = y_i$ for all i. Second, Bob mixes r with random elements $g_1, \cdots, g_{m+1} \in G_k$ belonging to the kernel of the homomorphism f_k and outputs the element $g = g_1 r_1 g_2 \cdots g_m r_m g_{m+1}$ as the ciphertext of h. Alice being able to compute f_k efficiently performs the decoding as follows:

$$f_k(g) = f_k(g_1 r_1 g_2 \cdots g_m r_m g_{m+1}) = f_k(r_1) \cdots f_k(r_m) = y_1 \cdots y_m = h.$$

The key point of such a system is to choose a presentation of the group G_k and the epimorphism f_k in order to provide the inverse of f_k to be a trapdoor function. The exact definition of homomorphic public-key cryptosystems and a survey of constructions can be found in [**GP, GP04**].

One way to implement the general concept of a homomorphic cryptosystem is to take G_k to be a subgroup of a certain group F such that the group operations in F can be performed in time polynomial in the size of operands. In the cryptosystems from [**GP**] and [**GP04**] the group F was taken as a free product of abelian groups and a modular group, respectively. In these cryptosystems the restriction of the mapping f_k to the set R_k was known publically and one can produce efficiently random $k^{O(1)}$-size elements of the group $\ker(f_k)$. In fact, the security of these cryptosystems was based on the difficulty of the membership problem (see below) for special subgroups of the group G_k.

Suppose first that the order of the group H_k is at most $k^{O(1)}$ (e.g. such an assumption was used in [**GP**]). Then using the generator set Y_k of H_k one can list all the elements $h_1, \ldots, h_m$ of this group in time $k^{O(1)}$ and then find within the same time a set $\{g_1, \ldots, g_m\}$ of distinct representatives of right cosets of $G_k = \ker(f_k)$ in G_k (one can set $g_i = f_k^{-1}(h_i)$ for all i). Now if an adversary Charlie could recognize efficiently the elements of G belonging to G_k, then he would efficiently compute $f_k(g)$ for all $g \in G_k$ due to the formulae

$$f_k(g) = f_k(g_i) \Leftrightarrow gg_i^{-1} \in G_k$$

where $i \in \{1, \ldots, m\}$. Thus, in this case the security of our cryptosystem is based on the intractability of the following problem:

Membership Testing (MT). *Given a group F and its subgroup G test whether a given $g \in F$ belongs to G.*

Suppose now the order of $H = H_k$ to be arbitrary. Then a quite natural way to break the cryptosystem is to find an expression of any $g \in G_k$ in the terms of generators belonging to the set X_σ (the attack of this kind was considered in [**GP04**]). Indeed, if Charlie could find efficiently for any element $g \in G_k$ an expression $g = x_1 \cdots x_m$ where $x_i \in X_k^{\pm}$ for all i, then he would efficiently compute $f_k(g)$ due to the formulae

$$f_k(g) = f_k(x_1) \cdots f_k(x_m) = f(x_1) \cdots f(x_m)$$

(we recall that the bijection $f : X_k \to Y_k$ is given publically). Thus, in this case we come to the presentation problem (see [**GP04**]). The MT problem and the presentation problem are closely related to each other (but generally could be not polynomial-time equivalent) and one can combine them in the following well-known problem of computational group theory (see [**BB93**]).

Constructive Membership Testing (CMT). *Given a group F and its subgroup G generated by a set X find an expression of a given $g \in F$ as a word in X, or determine that $g \notin G$.*

Last two decades a great attention was paid to CMT with different presentations of the group G. For example, if F is a subgroup of the symmetric group of degree $n \geq 1$, then the CMT can be solved in time $n^{O(1)}$ by the sift algorithm (see e.g. [**L93**]). In the case of groups $F = \mathrm{GL}(n, \mathbb{F})$ where $\mathbb{F}$ is an algebraic number field, there exists an effective Las Vegas algorithm solving CMT [**BB93**]. However, for $n = 1$ and $\mathbb{F}$ being a finite field, CMT is nothing else but the the Discrete Logarithm Problem. In [**BB93**] it was conjectured that CMT is difficult whenever the group G either involves a large abelian group as a quotient of a normal subgroup or has nonabelian composition factors which require large degree permutation representations. Finally, the problem becomes much more difficult if we take $F = \mathrm{GL}(n, R)$ to be the group of $n \times n$ invertible matrices over a ring R. In this case the problem is undecidable for $n = 4$ and $R = \mathbb{Z}$ (see [**M58**]).

3. Cryptographical generation of groups

3.1. A general scheme. We begin with a general scheme to construct a vast family of groups and homomorphisms supporting both key agreement protocols of Section 1 and homomorphic cryptosystems of Section 2. Let $\mathcal{G}$ be a class of groups closed with respect to a set $\mathcal{O}$ of group-theoretical operations of different arities (like direct or wreath products). For an integer $s \geq 1$ we denote by $\mathcal{O}_s$ a set of all operations of arity s belonging to $\mathcal{O}$. For a set $\mathcal{G}_0 \subset \mathcal{G}$ (playing the role of a starting family of groups in the construction) we define recursively a class $\mathcal{P}(\mathcal{G}_0, \mathcal{O})$ of pairs (G, T) where $G \in \mathcal{G}$ and T is a rooted labeled tree, as follows:

Base of recursion: any pair (G, T) with $G \in \mathcal{G}_0$ and T being the one-point tree with root labeled by G, belongs to $\mathcal{P}(\mathcal{G}_0, \mathcal{O})$.

Recursive step: given pairs $(G_1, T_1), \ldots, (G_s, T_s) \in \mathcal{P}(\mathcal{G}_0, \mathcal{O})$ and an operation $o \in \mathcal{O}_s$, the class $\mathcal{P}(\mathcal{G}_0, \mathcal{O})$ contains the pair (G, T) where $G = o(G_1, \ldots, G_s)$ and T is the tree obtained from $T_1, \ldots, T_s$ by adding a new root labeled by o and its sons being the roots of $T_1, \ldots, T_s$.

Let $(G, T) \in \mathcal{P}(\mathcal{G}_0, \mathcal{O})$. Then obviously $G \in \mathcal{G}$ and the *derivation tree* T of G provides the constructive proof for this membership. The group G is uniquely determined by T and we call it the *group associated with* T. The fact, that a derivation tree is an ordinary rooted tree the leaves and the internal vertices of which are labeled by elements of $\mathcal{G}_0$ and $\mathcal{O}$, respectively, enables us to choose a random derivation tree of a fixed size.

Suppose from now on that all the groups of $\mathcal{G}$ are given in a certain way (e.g., one can take as $\mathcal{G}$ a class of matrix groups given by generator sets). We assume also that for each operation $o \in \mathcal{O}_s$ and groups $G_1, \ldots, G_s \in \mathcal{G}$, the size $L(G)$ of the presentation of the group $G = o(G_1, \ldots, G_s)$ is at most $O(L)$ where $L = \sum_{i=1}^{s} L(G_i)$ and the group G can be constructed from $G_1, \ldots, G_s$ in time $L^{O(1)}$.

REMARK 3.1. *Thus the set of generators of G is assumed to be efficiently constructed; for instance, in the case of semidirect products (including both direct and wreath products considered below), this set is obtained by means of union of the generator sets of the operands.*

Let us define a size $L(T)$ of a derivation tree T to be the sum of the sizes of all labels of T; thus $L(T)$ includes the sizes of the groups assigned to the leaves of T together with the number of edges of T. Then the size of any pair $(G, T) \in \mathcal{P}(\mathcal{G}_0, \mathcal{O})$ is $O(L(T))$, and the knowledge of T enables us to find G in time polynomial in $L(T)$.

One of the problems arising in constructions of group-theoretical public-key cryptosystems is to find an efficient algorithm to produce a random group (or a collection of groups) belonging to a special class $\mathcal{G}$ and with a given size L of the presentation. Such a group G must be equipped with a private key providing an efficient solution of a certain computational problem for G that is supposedly difficult in the class $\mathcal{G}$ without knowledge of a private key. Our approach to the above problem is to choose an appropriate class $\mathcal{G}_0$ of groups, a set $\mathcal{O}$ of group-theoretical operations, and then to generate instances for the cryptosystem in question as follows:

Step 1: given a security parameter L choose randomly groups $G_1, \ldots, G_t \in \mathcal{G}_0$, such that $\sum_{i=1}^{t} L(G_i) = O(L)$;

Step 2: choose randomly a rooted labeled tree T of size $O(L)$ and with t leaves being labeled by $G_1, \ldots, G_t$;

Step 3: compute the group G associated with T (i.e. $(G, T) \in \mathcal{P}(\mathcal{G}_0, \mathcal{O})$);

Step 4: output the group G as a public key and the labeled tree T as a secret key.

Denote by $\mathcal{G}^*$ the class of groups G such that $(G, T) \in \mathcal{P}(\mathcal{G}_0, \mathcal{O})$ for some labeled tree T. Then the secrecy of the key T is based on the intractability of the following problem: given $G \in \mathcal{G}^*$ find a derivation tree T associated with G. A special case of this problem will be considered in Section 3.3.

For a homomorphic cryptosystem the above scheme is not sufficient because together with the group G we have to provide a group H and a secret homomorphism $f : G \to H$. To this end suppose that each group $G \in \mathcal{G}_0$ is equipped with a set $M(G)$ of homomorphisms $f : G \to H$ with $H \in \mathcal{G}_0$. We also assume that the following property holds:

Compatibility. For any operation $o \in \mathcal{O}_s$ and groups $G_i, H_i \in \mathcal{G}^*$, $i = 1, \ldots, s$, one can efficiently construct monomorphisms $\eta_i : G_i \to G$ and $\xi_i : H_i \to H$ where $G = o(G_1, \ldots, G_s)$ and $H = o(H_1, \ldots, H_s)$ such that given epimorphisms

$f_i : G_i \to H_i$ there exists an efficiently computed homomorphism $f : G \to H$ for which the equality $f \circ \eta_i = \xi_i \circ f_i$ holds for all i.

The constructed homomorphism is denoted by $o(f_1, \ldots, f_s)$. In this notation the set $M(\mathcal{G}_0, \mathcal{O})$ of instances f for a homomorphic cryptosystem can be defined recursively as follows:

Base of recursion: $M(G) \subset M(\mathcal{G}_0, \mathcal{O})$ for all $G \in \mathcal{G}_0$.

Recursion step: $o(f_1, \ldots, f_s) \in M(\mathcal{G}_0, \mathcal{O})$ for all $o \in \mathcal{O}_s$ and $f_1, \ldots, f_s \in M(\mathcal{G}_0, \mathcal{O})$.

We observe, that in the process of constructing the homomorphism $f : G \to H$ we also produce the derivation trees of the groups G and H. Obviously, these trees are isomorphic as unlabelled trees. We associate with f its derivation tree $\vec{T}$ which is constructed in a similar way as the derivation tree T of G. In fact, $\vec{T}$ is obtained from T by changing the labels of its leaves: a leaf of T with a label G_0 gets the label $f_0 \in M(G_0)$ corresponding to the choice of the homomorphism in the base of recursion.

Concerning a presentation of constructed homomorphisms we need to guarantee that properties (H1), (H2) and (H3) hold. To this end we assume that they hold for homomorphisms belonging to $M(G)$ for all $G \in \mathcal{G}_0$ and that they are preserved by operations from $\mathcal{O}$.

A realization of the exposed general schemes in finite matrix groups will be considered in the next subsection.

3.2. Generating matrix groups. Let us define the classes $\mathcal{G}_0, \mathcal{G}$ of groups and the set $\mathcal{O}$ of operations. First, we set

$$\mathcal{G} = \bigcup_{n,R} \{G : G \text{ is a subgroup of } \mathrm{GL}(n, R)\}$$

where n and R run over natural numbers and finite commutative rings, respectively. Thus, any $G \in \mathcal{G}$ is a group of $n \times n$ invertible matrices with entries belonging to R for some $n \in \mathbb{N}$ and some finite commutative ring R. We recall that any such ring is a direct sum of local commutative rings and each of the latter can be described via appropriate Galois ring: the Galois ring $\mathrm{GR}(p^m, r)$ of characteristic p^m and rank r is $\mathbb{Z}_{p^m}[x]/(f)$ where $f \in \mathbb{Z}_{p^m}[x]$ is a monic polynomial of degree r whose image in $\mathbb{Z}_p[x]$ is irreducible (see [**MD74**]). We note that $\mathrm{GR}(p^m, r)$ is a local ring whose radical $\mathrm{Rad}(\mathrm{GR}(p^m, r))$ equals to (p).

PROPOSITION 3.2. [**MD74, W03**] *Let R be a finite commutative local ring of characteristic p^m and $\mathbb{F} = \mathrm{GF}(p^r)$ the residue field of R. Then*

(1) *$R^\times = \mathcal{T} \times (1_R + \mathrm{Rad}(R))$ where $\mathcal{T}$ is a cyclic group isomorphic to $\mathbb{F}^\times$,*
(2) *the subring R_0 of R generated by $\mathcal{T}$ is a Galois ring $\mathrm{GR}(p^m, r)$,*
(3) *R is a homomorphic image of the ring $R_0[X_1, \ldots, X_t]$ where t is the minimal size of a generator set of the radical of R.* ∎

PROPOSITION 3.3. [**MD74**] *Let p be a prime and m, r be natural numbers. Then*

(1) *there exists the unique up to isomorphism Galois ring $\mathrm{GR}(p^m, r)$ of characteristic p^m and rank r,*

(2) *each element x of $\mathrm{GR}(p^m, r)$ is uniquely represented in the form $x = \sum_{i=0}^{m-1} t_i p^i$ where $t_i \in \mathcal{T} \cup \{0\}$ for all i,*
(3) *given $\bar{\sigma} \in \mathrm{Aut}(\mathbb{F})$ the mapping $x \mapsto \sum_{i=0}^{m-1} t_i^\sigma p^i$ where σ is the automorphism of the group $\mathcal{T}$ induced by $\bar{\sigma}$ (see statement (1) of Proposition 3.2), is an automorphism of $\mathrm{GR}(p^m, r)$.* ∎

To construct a pool of finite commutative rings R one can start with the ring $R = \mathbb{Z}_m$ (as the recursion base) and to extend it repeatedly, for example, by taking of:

(R1) the full matrix ring $\mathrm{Mat}(n, R)$ for some integer $n \geq 2$,
(R2) the group ring $R[G]$ for a finite commutative group G,
(R3) the quotient ring $R[X]/(\lambda)$ for a univariate polynomial $\lambda \in R[X]$.

In particular, construction (R3) produces all the Galois rings. We also remark that since the factorization of the characteristic of the resulting ring is not given, the decomposition in local summands is not presented explicitly.

We define a set $\mathcal{G}_0 \subset \mathcal{G}$ to be a class of classical simple (including abelian) subgroups G of the groups $\mathrm{GL}(n, \mathbb{F})$ where $n \in \mathbb{N}$ and $\mathbb{F}$ is a finite field. Any such group $G \in \mathcal{G}_0$ is given by a set of generators; for an abelian group of a prime order p one can use, e.g. its two-dimensional representation

$$(3) \qquad \mathbb{Z}_p^+ \to \mathrm{GL}(2, p), \quad x \mapsto \begin{pmatrix} 1 & x \\ 0 & 1 \end{pmatrix}$$

In fact, it is not necessary that $\mathcal{G}_0$ contains all classical groups; one can form $\mathcal{G}_0$ from the group of special types, e.g. $\mathrm{PSL}(n, \mathbb{F})$ or something like that. Since the elements of $\mathcal{G}_0$ are parametrized by the tuples of natural numbers, one can efficiently choose a random group $G \in \mathcal{G}_0$ with a given size $L(G)$ of presentation.

The choice of the set $\mathcal{O}$ of operations was inspired by the Aschbacher theorem [**A84**] on classifying maximal subgroups of classical groups. Let us describe these operations.

Changing the underlying ring. Let R be a finite commutative ring and R' be an extension of R. Then the natural monomorphism

$$\varphi : \mathrm{GL}(n, R) \to \mathrm{GL}(n, R')$$

gives an unary operation in $\mathcal{G}$ taking $G \in \mathcal{G}$ to $\varphi(G)$. This operation can be performed efficiently whenever e.g. the embedding R to R' is given explicitly and the number $d = [(R')^+ : R^+]$ is small. As possible constructions of extensions we suggest the extension of $\mathbb{Z}_m$ to $\mathbb{Z}_{m'}$ where m is a divisor of m', and the ones described above in (R1), (R2) and (R3).

Conversely, any embedding of the ring R' into the ring $\mathrm{Mat}(d, R)$ induces the natural monomorphism

$$\varphi' : \mathrm{GL}(n, R') \to \mathrm{GL}(nd, R)$$

taking a matrix of $\mathrm{GL}(n, R')$ to the block matrix of $\mathrm{GL}(nd, R)$ with d^2 blocks of size n. As possible constructions of embeddings we suggest the natural embedding of a field of the order q^d to $\mathrm{Mat}(d, q)$, or the direct sum of d copies of R to $\mathrm{Mat}(d, R)$, or the natural embeddings arising from constructions (R1), (R2) and (R3). This produces another unary operation in $\mathcal{G}$ taking $G \in \mathcal{G}$ to $\varphi'(G)$. In order not to blow up the size of representation one should assume that d is small.

In both cases the isomorphism type of the group G (as an abstract group) is not changed, but the operations change it as a linear group.

Direct products. Suppose that groups $G_1, \ldots, G_s \in \mathcal{G}$ are such that $G_i \leq \mathrm{GL}(n_i, R)$ where $n_i \in \mathbb{N}$ and R is a finite commutative ring. Then

$$G = G_1 \otimes \cdots \otimes G_s \leq \mathrm{GL}(n, R)$$

where $n = \prod_{i=1}^{s} n_i$, and we obtain an s-ary operation in $\mathcal{G}$. A set of generators for the group G can be efficiently constructed from the generating sets for $G_1, \ldots, G_s$ by means of the Kronecker product of the corresponding matrices. When R is a field the group G is irreducible iff so are the groups $G_1, \ldots, G_s$. (A matrix group G is called irreducible if the underlying linear space contains no nontrivial G-invariant subspaces.)

Similarly, if $m = n_i$, $G_i \cap G_i' = \{I_m\}$ and G_i' normalizes G_i for all $i = 1, \ldots, s$ where G_i' is the group generated by G_j, $j \neq i$, then $G_1 \times \cdots \times G_s$ is a subgroup of $\mathrm{GL}(m, R)$ which gives one more s-ary operation.

Wreath products. The *wreath product* $G \wr \Gamma$ of a group G and a permutation group $\Gamma \leq \mathrm{Sym}(m)$ is defined to be the semidirect product of the m-fold direct product $G^m = G \times \cdots \times G$ by the group Γ acting on G^m via coordinatewise permutations. If $G \leq \mathrm{GL}(n, R)$, then the group $G \wr \Gamma$ has two natural linear representations obtained from the natural monomorphisms

$$G^m \to \mathrm{GL}(nm, R), \qquad G^m \to \mathrm{GL}(n^m, R),$$

the first of which is induced by the m-fold direct sum of the underlying R-module, whereas the second one is induced by the m-fold tensor product of it. The images of the group $G \wr \Gamma$ under these two monomorphisms are called the *imprimitive* and the *product actions* of the wreath product, respectively. Thus, we obtain two more efficiently computable unary operations in $\mathcal{G}$ for each permutation group Γ. For our purpose it is enough to set Γ to be the symmetric group $\mathrm{Sym}(m)$. (More elaborated way could be based on the fact that any transitive group is obtained from the action of a group on the set of right cosets of some subgroup by means of right multiplications.) In the case of R being a field the resulting groups are always irreducible whenever G is irreducible and Γ is transitive.

Conjugations. An obvious unary operation in $\mathcal{G}$ consists in the conjugation of a group $G \subset \mathrm{GL}(n, R)$ by means of a randomly chosen matrix from $\mathrm{GL}(n, R)$. Such an operation enables us to hide the form of a generator set of the group G.

Let $\mathcal{O}$ be the set of the above operations and $\mathcal{G}^* \subset \mathcal{G}$ be the set of all groups G such that $(G, T) \in \mathcal{P}(\mathcal{G}_0, \mathcal{O})$ for some rooted labeled tree T (see Subsection 3.1). In the following statement we consider the specializations of the problems MT (see Section 2) and LTP (see Subsection 1.3) for the class $\mathcal{G}^*$. In both cases we suppose that the group $G \in \mathcal{G}^*$ is given by a set of generators. If $G \leq \mathrm{GL}(n, R)$ for a certain $n \in \mathbb{N}$ and for a finite commutative ring R, then in the case of LTP we set V to be the standard free R-module of dimension n on which the group $\mathrm{GL}(n, R)$ acts, whereas for MT problem we set $F = \mathrm{GL}(n, R)$.

LEMMA 3.4. *Let $G \in \mathcal{G}^*$. Then given a derivation tree of G the problems MT and LTP can be solved in time polynomial in $L(G)$.*

Proof. Let T be a derivation tree of G. Then the labels of the leaves of T are the groups $G_1, \ldots, G_t \in \mathcal{G}_0$. Due to the choice of $\mathcal{G}_0$ the problems MT and LTP can be solved for the group G_i in time polynomial in $L(G_i)$ for $i = 1, \ldots, t$. (Indeed, any nonabelian classical matrix group is given together with a suitable matrix representation which can be used for testing membership; for an abelian group representation (3) provides a trivial membership testing algorithm).

Since $L(G) = L(T)^{O(1)}$, it suffices to verify that by means of the construction of the tree T the problems can be reduced in time $L(T)^{O(1)}$ to the corresponding problems for $G_1, \ldots, G_t$. For this purpose let us consider, for instance, the reduction in the case of the primitive wreath product $G = H \wr \Gamma$ with $H \leq \mathrm{GL}(n, R)$ and $\Gamma = \mathrm{Sym}(m)$ (other operations from $\mathcal{O}$ on groups are treated in a similar way). Then $G \leq \mathrm{GL}(n^m, R)$ and since T is given, we know the decomposition

$$V = U \otimes \cdots \otimes U \quad (m \text{ times})$$

where V and U are the standard R-modules for groups $\mathrm{GL}(n^m, R)$ and $\mathrm{GL}(n, R)$, respectively. Any element $g \in G$ can be represented as a pair $(h, k) \in H^m \times \mathrm{Sym}(m)$ such that

(4) $$(u_1, \ldots, u_m)^g = (u_{i_1}^{h_{i_1}}, \ldots, u_{i_m}^{h_{i_m}})$$

where $h = (h_1, \ldots, h_m)$ and $i_j = j^{k^{-1}}$ for $j = 1, \ldots, m$. Now the permutation k can be efficiently computed from the elements of the form $(0_R, \ldots, 1_R, \ldots, 0_R)^g$ (with 1_R being the unique nonzero component in a certain place). So, the element $h = gg_k^{-1}$ also can be found efficiently where g_k is the element of $\mathrm{GL}(V) = \mathrm{GL}(n^m, R)$ corresponding to k (this element acts on V exactly by permuting coordinates according to k). In particular, this provides a polynomial time reduction of the MT problem for G to the corresponding problem for H.

Next, proceeding to the LTP problem let $v \in u^G$ for some $u, v \in V$. Denote by D the bipartite graph with parts being the multisets $\{u_1, \ldots, u_m\}$ and $\{v_1, \ldots, v_m\}$ and the edges being the pairs (u_i, v_j) for which $v_i \in (u_j)^H$. Then from (4) it follows that there is a one to one correspondence between the matchings $\{(u_i, v_{j_i}) : i = 1, \ldots, m\}$ of the graph D and the set $\{k \in \Gamma : v = u^g \text{ with } g = (h, k) \in G \text{ for some } h \in H^m\}$. Since the problem of finding a matching of a bipartite graph can be solved efficiently, we see that the LTP problem for G is polynomial time reducible to the corresponding problem for H. ∎

A natural way to apply our construction to the key agreement protocol is to choose a random group $G \in \mathcal{G}^*$ of a prescribed size and then choose random subgroups G_A and G_B of G (see (1)). These groups can be specified by sets of generators constructed as follows:

Step 1. Let $\mathcal{L}$ be the set of leaves of the derivation tree T of the group G. For each $l \in \mathcal{L}$ take random subsets $X_A(l)$ and $X_B(l)$ of the group H_l associated with l.

Step 2. Construct the trees T_A and T_B obtained from T only relabelling of the leaves: the leaf $l \in \mathcal{L}$ is labeled by the group $H_A(l) = \langle X_A(l) \rangle$ in T_A and by the group $H_B(l) = \langle X_B(l) \rangle$ in T_B respectively.

Step 3. Set G_A and G_B to be the groups the derivation trees of which are T_A and T_B (the generator sets X_A of G_A and X_B of G_B are obtained in accordance with Remark 3.1 on constructing the generators).

Thus, the constructing of the groups G_A and G_B is performed simultaneously with the constructing the group G. (In fact, all we need, is the embedding of each group assigned to a leaf of the derivation tree of the group G into G.) In this way it is possible to control some properties of the groups, for instance, to avoid the situation when G_A centralizes G_B (then the common key coincides with 1_G and so is not secure).

Applying our construction to design homomorphic cryptosystems is more delicate. First of all we define the set $M(G)$ for each group $G \leq \mathrm{GL}(n, R)$ for some $n \in \mathbb{N}$ and some finite commutative ring R (note that this covers the case $G \in \mathcal{G}_0$ and also allows one to produce homomorphisms in one more way: replacing $\mathcal{G}_0$ by a bigger subclass of $\mathcal{G}$). Namely, any automorphism $\sigma \in \mathrm{Aut}(R)$ induces a homomorphism

$$f_\sigma : G \to G^\sigma,\ A \mapsto A^\sigma$$

where the matrix $A^\sigma \in \mathrm{GL}(n, R)$ is obtained from the matrix $A \in \mathrm{GL}(n, R)$ by entry-wise applying of σ. To choose σ we observe that $R = \oplus_{i \in I} R_i$ where each R_i is a finite local commutative ring. Any automorphism of the residue field of the ring R_i can be lifted to the automorphism of this ring (statement (3) of Proposition 3.3). In the representation of the Galois ring as a quotient ring of a ring of polynomials this lifting can be done efficiently. Taking any collection $\{\sigma_i\}_{i \in I}$ one can construct the automorphism $\sigma \in \mathrm{Aut}(R)$ such that $\sigma|_{R_i} = \sigma_i$ for all i. The set of such automorphisms we denote by $\mathrm{Aut}_0(R)$ (in the case of R being a field this group coincides with $\mathrm{Aut}(R)$). Set

(5) $$M(G) = f_0 \cup \{f_\sigma :\ \sigma \in \mathrm{Aut}_0(R)\}$$

where f_0 is a trivial homomorphism taking any element of G to the identity matrix of $\mathrm{GL}(n, R)$. We assume that each $f \in M(G)$ is given by the images of generators of the group G and hence conditions (H1), (H2) hold trivially, while condition (H3) is obvious for f_σ and follows from the choice of presentation of G (by generators) for f_0. Then assuming that the ring R is given explicitly, one can choose a random element of $M(G)$ in time polynomial in $L(G)$.

To provide the recursive step in constructing a homomorphism $f \in M(\mathcal{G}_0, \mathcal{O})$ it suffices to verify the compatibility property for the set $\mathcal{O}$ of the operations (see Subsection 3.1) and to verify that the operations preserve properties (H2) and (H3) (see Section 2). However, the compatibility property is obviously fulfilled if the required monomorphisms η_i for f are chosen to be

- identical in case of the operation *changing the underlying ring*,
- the embedding $g_i \mapsto 1_{G_1} \otimes \cdots \otimes g_i \otimes \cdots \otimes 1_{G_s}$ in case of the operation *direct product*,
- the embedding $g \mapsto (g, \ldots g; 1_\Gamma)$ in case of the operation *wreath product*,
- the isomorphism $g \mapsto a^{-1}ga$ in case of the operation of *conjugation* by a.

(The monomorphisms ξ_i are defined in a similar way.)

Concerning properties (H2) and (H3) we note that they are obvious for the operations changing the underlying ring and conjugation. In the case of the direct product it suffices to note that a generator $1_{G_1} \otimes \cdots \otimes g_i \otimes \cdots \otimes 1_{G_s}$ of the group G (where g_i is a generator of G_i) is mapped under f to $1_{G_1} \otimes \cdots \otimes f_i(g_i) \otimes \cdots \otimes 1_{G_s}$, and $\ker(f) = \ker(f_1) \otimes \cdots \otimes \ker(f_s)$. In the case of the wreath product generators $(1_{G_1}, \ldots, g_i, \ldots, 1_{G_s}; 1_\Gamma)$ and $(1_{G_1}, \ldots, 1_{G_s}; \gamma)$ (where γ is a generator of Γ) of the group G are mapped under f to $(1_{G_1}, \ldots, f_i(g_i), \ldots, 1_{G_s}; 1_\Gamma)$ and to $(1_{G_1}, \ldots, 1_{G_s}; \gamma)$, respectively, and finally $\ker(f) = (\ker(f_1), \cdots, \ker(f_s); 1_\Gamma)$.

Thus, in all the cases, the resulting homomorphism is efficiently computable. The above discussion shows that the following statement holds.

LEMMA 3.5. *Let $f : G \to H$ be a homomorphism constructed in the above way where $G, H \in \mathcal{G}^*$. Then given a derivation tree $\overrightarrow{T}$ of f (see the end of Subsection 3.1) one can find $f(g)$ for $g \in G$ in time polynomial in $L(G)$ and in the size of g.* ∎

3.3. Secure generation. Let us fix the classes $\mathcal{G}_0, \mathcal{G}, \mathcal{G}^*$, the set $\mathcal{O}$ of operations and the sets $M(G)$ for $G \in \mathcal{G}_0$ as in Subsection 3.2. Then due to Lemmas 3.4 (resp. Lemma 3.5) one can construct groups $G \in \mathcal{G}^*$ (resp. homomorphisms $f \in M(\mathcal{G}_0, \mathcal{O})$) to realize key agreement protocols (resp. homomorphic cryptosystems) in which the group G (resp. the homomorphism f) and the derivation tree T of G (resp. $\overrightarrow{T}$ of f) play the roles of public and secret keys, correspondingly. The security of such systems is based on the difficulty of the following problem.

Decomposition Problem. *Given a group $G \in \mathcal{G}^*$ (resp. a homomorphism $f \in M(\mathcal{G}_0, \mathcal{O})$) find a derivation tree T of G (resp. $\overrightarrow{T}$ of f).*

This problem arises in connection with a computational version of the above mentioned Aschbacher's theorem. A number of practical algorithms (without complexity bounds) for Decomposition Problem are known (see [**LG01**]), but in general this problem seems to be difficult. Indeed, suppose that $R = \mathbb{Z}_m$ where $m = pq$ with p and q being two different primes. Denote by G_p the cyclic matrix group of the order p in $\mathrm{GL}(2, p)$ (see (3)). Similarly, the group G_q is defined. Then $G_p, G_q \in \mathcal{G}_0$ and
$$G = G'_p \times G'_q \leq \mathrm{GL}(2, R)$$
where G'_p and G'_q are the images of the groups G_p and G_q with respect to the natural embeddings $\mathrm{GL}(2, p)$ and $\mathrm{GL}(2, q)$ into $\mathrm{GL}(2, R)$. Thus, the group G can be constructed in two steps: first one constructs the groups G'_p and G'_q (the operation of changing the underlying ring), and then one sets $G = G'_1 \times G'_2$ (the operation of the direct product). This implies that $G \in \mathcal{G}^*$. This shows that the integer factoring problem is a special case of the Decomposition Problem.

Another strategy of an adversary Charlie could be to avoid solving the Decomposition Problem and to try to solve the problems like LTP, SCSP or CMT directly. To prevent such an attack one can choose the leaves of a derivation tree of the group G to be the groups of the size exponential with respect to $L(G)$. Then from the construction it follows that these groups will appear as the composition factors of G. However, for the groups with large composition factors all the problems like LTP, SCSP or CMT seem to be difficult (cf. Subsection 1.3 and Section 2).

We mention one more attack of Charlie for the case of a homomorphic cryptosystem. Suppose we construct in the above way the homomorphism $f : G \to H$ with $G, H \in \mathcal{G}^*$. We call the homomorphism *linear* if it induces the ring homomorphism $f' : A(G) \to A(H)$ where $A(G)$ (resp. $A(H)$) is the subring of the underlying full matrix ring generated by G (resp. H). For a linear homomorphism the corresponding homomorphic cryptosystem can be easily broken whenever $G \leq \mathrm{GL}(n, R)$ where $R = \mathbb{Z}_n$ for some $n \in \mathbb{N}$ or R is a finite field (or, more generally, a direct sum of Galois rings). Indeed, in this case Charlie can find $f(g)$ for $g \in G$ as follows. Take random generators $g_1, \ldots, g_s$ of the group G and find a decomposition $g = \sum_{i=1}^{s} c_i g_i$ with $c_i \in R$ just involving linear algebra. Then $f(g) = \sum_{i=1}^{s} c_i f(g_i)$ due to the linearity of f. To prevent this attack one can take some initial homomorphisms at the leaves of the derivation tree to be elements of the group $\text{Aut}_0(R)$ (see (5)). Then the constructed homomorphism is not linear in general (e.g. if $g \in \text{GL}(n, \mathbb{F})$ with $\mathbb{F}$ being a field, and $\sigma \in \text{Aut}(\mathbb{F})$, then generally $(ag)^\sigma \neq ag^\sigma$).

We conjecture that two-party key agreement protocol and homomorphic public-key cryptosystem based on the constructed class of matrix groups over finite commutative rings are secure (as mass problems). If the latter was true, then one could construct encrypted simulation of a boolean circuit of the logarithmic depth (the details can be found in [**GP**]).

Final remarks

One of the main problems in constructing homomorphic public-key cryptosystems consists in finding appropriate trapdoor functions. However, in the natural presentations of homomorphisms of algebraic structures the problem of breaking such a system is reduced to some variants of the CMT problem. On the other hand, there is the following result for matrix groups over finite fields.

THEOREM 3.6. [**KS03**, Theorem 6.1] *Given $K = \langle X \rangle \leq \text{GL}(d, p^e)$ where $X \subset \text{GL}(d, p^e)$, there is a Las Vegas algorithm that given any $g \in \text{GL}(d, p^e)$, decides whether $g \in K$, and if $g \in K$, then the algorithm produces a straight-line program with the input X, yielding g. The algorithm uses an oracle to compute discrete logarithms in fields of characteristic p with sizes up to p^{ed}. In case when all of those composition factors of Lie type in characteristic p are constructively recognizable with a Discrete Log oracle* [1], *the running time is a polynomial in the input length $|X|d^2 e \log p$, plus the time required for polynomially many calls to the Discrete Log oracle.* ∎

This theorem shows that having an oracle for the Discrete Logarithm, the membership problem can be solved in probabilistic polynomial time for matrix groups over finite fields. This means that at least for homomorphic public-key cryptosystems over such groups there is a little hope to find a trapdoor function different from functions the difficulty of inversion of which is based on the intractability of the Discrete Logarithm. However, only a little is known on the computational complexity of the membership problem for matrix groups over rings. So, constructions over such groups seem to be more perspective from the point of view of algebraic (non-commutative) cryptography.

References

[AAG99] I. Anshel, M. Anshel, D. Goldfeld, *An algebraic method for public-key cryptography*, Mathematical Research Letters, **6** (1999), 287–291.

[A84] M. Aschbacher, *On the maximal subgroups of the finite classical groups*, Invent. Math., **76** (1984), 469–514.

[BB93] R. Beals, L. Babai, *Las Vegas algorithms for matrix groups*, Proc. 34th IEEE FOCS, 1993, 427–436.

---

[1]The current list of groups of Lie type recognizable with a Discrete Log oracle is given in [**KS03**]; this list includes the groups of series A, B, C, D.

[CY97] A. Chistov, G. Ivanyos, M. Karpinski, *Polynomial time algorithms for modules over finite dimensional algebras*, Proceedings of the 1997 International Symposium on Symbolic and Algebraic Computation (Kihei, HI), 68–74, ACM, New York, 1997.

[VJS88] Do Long Van, A. Jeyanthi, R. Siromoney, K. G. Subramanian, *Public key cryptosystem based on word problems*, ICOMIDC Symp. Math. of Computation, Ho Chi Minh City, April, 1988.

[EK] B. Eick, D. Kahrobaei, *Polycyclic groups: a new platform for cryptology?*, arXiv.math.-GR/0411077.

[FM91] J. Feigenbaum, M. Merritt, *Open questions, talk abstracts, and summary of discussions*, DIMACS series in discrete mathematics and theoretical computer science, **2** (1991), 1–45.

[GB01] S. Goldwasser, M. Bellare, *Lecture Notes on Cryptography*, http://www-cse.ucsd.edu/users/mihir/papers/gb.html, 2001.

[GM84] S. Goldwasser. S. Micali, *Probabilistic encryption*, J.Comput.Syst.Sci., **28** (1984), 270–299.

[G05] D. Grigoriev, *Public-key cryptography and invariant theory*, J. Math. Sci., **126** (2005), 1152–1157.

[GP04] D. Grigoriev, I. Ponomarenko, *Homomorphic public-key cryptosystems over groups and rings*, Quaderni di Matematica, **13** (2004), 305–325.

[GP] D. Grigoriev, I. Ponomarenko, *Homomorphic public-key cryptosystems and encrypting boolean circuits*, to appear in Appl. Alg. Eng. Communic. Comput., 2006.

[KS03] W. M. Kantor, A. Seress, *Computing with matrix groups*, Groups, combinatorics & geometry (Durham, 2001), 123–137, World Sci. Publishing, River Edge, NJ, 2003.

[KLCHKP] K. H. Ko, S. J. Lee, J. H. Cheon, J. W. Han, J. Kang, C. Park, *New public-key cryptosystem using braid groups*, Lecture Notes in Computer Science, **1880** (2000), 166–183.

[LG01] C. R. Leedham-Green, *The computational matrix group project*, pp. 229–247 in: Groups and Computation III (eds. W. M. Kantor and A. Seress), The Ohio State Univ. Math. Res. Inst. Publ. 8, Walter deGruyter, BerlinNew York 2001.

[L93] E. M. Luks, *Permutation groups and polynomial-time computation*, DIMACS Series in Discrete Mathematics and Theoretical Computer Science, **11** (1993), 139–175.

[MKS66] W. Magnus, A. Karrass, D. Solitar, *Combinatorial group theory: Presentations of groups in terms of generators and relations*, Interscience Publishers, New York-London-Sydney, 1966.

[MD74] B. R. MacDonald, *Finite Rings with Identity*, New York, Marcel Dekker, 1974.

[M58] K. A. Mihailova, *The occurrence problem for a direct product of groups*, Dokl. Akad. Nauk. **119** (1958), 1103–1105. (in Russian)

[M99] T. Miyazaki, *Polynomial-time computation in matrix groups*, Ph.D. dissertation, Tech. Rep. CISTR9911, Department of Computer and Information Science, University of Oregon, Eugene, 1999.

[NS98] D. Naccache, J. Stern, *A new public-key cryptosystem based on higher residues*, Proc. 5th ACM Conference on Computer and Communication Security, 1998, 59–66.

[PKHK] S.-H. Paeng, D. Kwon, K.-C. Ha, J. H. Kim, *Improved public-key cryptosystem using finite non-abelian groups*, IACR ePrint 2001/066.

[RAD78] R. Rivest, L. Adleman, M. Dertouzos, *On data banks and privacy homomorphisms*, Found. of Secure Computations, Academic Press, 1978, 169–179.

[SU06] V. Shpilrain, A. Ushakov, *The conjugacy search problem in public-key cryptography: unnecessary and insufficient*, to appear in Appl. Alg. Eng. Communic. Comput., 2006.

[S76] D. A. Suprunenko, *Matrix groups*, AMS, Providence, 1976.

[W03] Z.-X. Wan, *Lectures on finite fields and Galois rings*, World Scientific Publishing Co., Inc., River Edge, NJ, 2003.

IRMAR, Université de Rennes, Beaulieu, 35042, Rennes, France
E-mail address: dmitry.grigoryev@univ-rennes1.fr

Petersburg Department of V.A.Steklov Institute of Mathematics, Fontanka 27, St. Petersburg 191023, Russia
E-mail address: inp@pdmi.ras.ru

A Practical Attack on the Root Problem in Braid Groups

Anja Groch, Dennis Hofheinz, and Rainer Steinwandt*

ABSTRACT. Using a simple heuristic approach to the *root problem* in braid groups, we show that cryptographic parameters proposed in this context must be considered as insecure. In our experiments we can, often within seconds, extract the secret key of an authentication system of Sibert et al. that is based on the root problem in braid groups.

1. Introduction

In recent years several proposals for asymmetric cryptographic schemes building on braid groups have been made. For an excellent introduction to the subject we refer to Dehornoy's survey article [**Deh04**]. Next to proposals based on (variants of) the *conjugacy problem*, more recently constructions for making use of the *root problem* in braid groups gained attention, and in the sequel we show that parameters proposed in this context succumb to a simple heuristic attack. Our approach reduces a root problem in the braid group B_n to a root problem in the symmetric group S_n, and despite its simplicity turns out as rather effective for proposed instances. To describe the details, we start by recalling some basic terminology on braid groups.

1.1. Braid Groups. For our purposes it suffices to interpret the *braid group* B_n ($n \in \mathbb{N}$) as a finitely presented group defined through the presentation (cf. [**Art25**])

$$\left\langle \sigma_1, \ldots, \sigma_{n-1} \;\middle|\; \begin{array}{rcll} \sigma_i \sigma_j \sigma_i & = & \sigma_j \sigma_i \sigma_j & \text{if } |i-j| = 1 \\ \sigma_i \sigma_j & = & \sigma_j \sigma_i & \text{if } |i-j| > 1 \end{array} \right\rangle.$$

As usual, we refer to $\sigma_1, \ldots, \sigma_{n-1}$ as *(Artin) generators* and to arbitrary elements of B_n as *braids*. To refer to a specific representation of a braid in terms of Artin generators, we use the term *braid word*. A braid is said to be *positive* if and only if it can be written as a product of generators σ_i, i.e., without involving negative powers of the σ_i. Here the identity $\varepsilon \in B_n$ is regarded as positive, too, and it can be shown that the positive braids in B_n form a *monoid* B_n^+ which embeds into B_n

---

1991 *Mathematics Subject Classification.* Primary 94A60; Secondary 20F36.
Key words and phrases. braid group, root problem, cryptanalysis.
*This work was partially supported by a Federal Earmark grant for *Research in Secure Telecommunication Networks (2005-06)*.

(cf. [**Gar69**]). Moreover, we call a braid $u \in B_n^+$ a *tail* of $w \in B_n^+$ if w can be written as $w = vu$ for some $v \in B_n^+$. Analogously, $u \in B_n^+$ is referred to as a *head* of $w \in B_n^+$ if $w = uv$ for some $v \in B_n^+$.

1.2. Δ-Normal Form. Setting inductively $\Delta_1 := \sigma_1$ and $\Delta_i = \sigma_1 \cdots \sigma_i \cdot \Delta_{i-1}$ for $1 < i < n$, we define the *fundamental braid* $\Delta \in B_n$ as $\Delta := \Delta_{n-1}$. Next, we establish a partial ordering $\leq$ on the elements of B_n: For $u, v \in B_n$ we set $v \leq w$ if and only if there exist positive braids $\alpha, \beta \in B_n^+$ such that $w = \alpha v \beta$. Now any braid $\alpha \in B_n$ satisfying $\varepsilon \leq \alpha \leq \Delta$ is referred to as *canonical factor*. Further on, we have a *canonical homomorphism* $\pi : B_n \longrightarrow S_n$ from the braid group B_n into the symmetric group S_n such that

- $\pi(\sigma_i) = (i, i+1)$ and
- restricting π to the set of canonical factors in B_n induces a bijection

(see [**EM94**]). We denote by π^{-1} the map that maps permutations to canonical factors such that $\pi \circ \pi^{-1} = \text{id}$.

A factorization $\gamma = \alpha\beta$ of a positive braid γ into a canonical factor α and a positive braid β is said to be *left-weighted* if and only if α has the maximal length among all such decompositions. A *right-weighted* factorization is defined analogously. Moreover, for any braid $w \in B_n$ we denote the greatest $i \in \mathbb{Z}$ with $\Delta^i \leq w$ by $\inf w$; analogously, $\sup w$ stands for the smallest $i \in \mathbb{Z}$ with $w \leq \Delta^i$. With this notation every braid $w \in B_n$ can be written *uniquely* as

$$(1) \qquad w = \Delta^r W_1 \cdots W_s$$

with $r = \inf w$, $s = \sup w - \inf w$ and canonical factors $\varepsilon < W_i < \Delta$ such that $W_i W_{i+1}$ is left-weighted for $1 \leq i < s$ (cf. [**Gar69, EM94**]). In this context, we say that s is the *canonical length* of w, and the explicit decomposition (1) is called the Δ-*normal form* of w.

Note that the W_i are canonical factors, so they can be represented uniquely by the corresponding permutations $\pi(W_i) \in S_n$. For a given braid word $w \in B_n$, its Δ-normal form can be computed in time $\mathcal{O}(|w|^2 n \log n)$ with $|\cdot|$ denoting the word length (see [**CKL$^+$01**] for details). Sometimes the Δ-normal from is called *left-normal form* because of the power of Δ being written on the left and the decompositions of canonical factors being left-weighted. Writing the power of Δ on the right and demanding right-weightedness of the decompositions of canonical factors, one obtains the *right-normal form*, which can also be computed efficiently.

1.3. Conjugacy and Root Problem. For the *conjugacy (search) problem* (short: *CSP*), we are given two braid words so that the corresponding braids $u, w \in B_n$ are related by $w = \alpha^{-1} u \alpha$ for some $\alpha \in B_n$, and the goal is to find some braid $\alpha' \in B_n$ with $w = \alpha'^{-1} u \alpha'$. Leaving aside the question of efficiency, this problem has been solved in [**Gar69**], but finding an efficient algorithm for CSP is the topic of ongoing research. In parts motivated by cryptanalytic purposes, a number of heuristic approaches to the CSP have been brought up (see, e.g., [**Hug02, HS02, GKT$^+$05**]). An experimentally very efficient Las Vegas algorithm for solving the CSP is known, too (cf. [**Geb03**]), and from the cryptographic point of view further research on how to generate hard instances of the CSP is desirable.

Most of the braid group based cryptographic schemes that have been proposed rely on the difficulty of some variation of the CSP. For instance, the public key cryptosystem of [**KLC$^+$00**] is based on a variant of the CSP, where one is given

$\alpha^{-1}u\alpha$ and $\beta^{-1}u\beta$ for α and β chosen from two commuting subgroups, and one needs to find $(\alpha\beta)^{-1}u(\alpha\beta)$ (see also [**SU04**]). Note that this particular problem can be solved in polynomial time, cf. [**CJ03**].

For the *root extraction problem* (short: *REP*), we are given a braid word representing some $w \in B_n$ along with some natural number $m \in \mathbb{N}$ such that $w = u^m$ for some $u \in B_n$. The goal is to find a braid u' with $u'^m = w$. As shown by González-Meneses [**GM03**], u' and u then in particular must be conjugate.

Using the REP for cryptography has been suggested by Ko et al. in [**KLC+00**], and a concrete cryptographic system for entity authentication basing on the hardness of the REP has been put forward by Sibert et al. in [**SDG02, Sib03**]. This system is provably secure (assuming the hardness of REP *and* CSP) and is described in more detail in Section 2. Very recently, in [**LC05**], two authentication schemes, claimed to be based on the REP, were suggested, but the first of these has been attacked successfully in [**Tsa05**] without solving the REP. The second scheme is unsatisfying in the sense that is relies on the difficulty of the Diffie-Hellman decomposition problem which can be solved in polynomial time [**CJ03**]. Moreover, [**LC05**] lacks concrete parameter suggestions that could serve as testground for practical tests. Hence here we do not investigate these two proposals further.

Disregarding the question of efficiency, a result of Styshnev [**Sty79**] provides an algorithmic solution for the REP. Furthermore, Lee in [**Lee04**] shows how to reduce the REP to a conjugacy search problem in a different (Garside) group. In combination with, e.g., the techniques developed in [**Geb03**] (which apply to arbitrary Garside groups) this can be used to solve the REP. However, to our knowledge no results about the efficiency of that method are currently known. In particular, it seems unclear how efficient the algorithm of [**Geb03**] performs in the Garside groups considered in [**Lee04**]. While exploring this question certainly looks worthwhile, in the sequel we show that already a rather simple heuristic approach enables a practical attack on proposed parameters of the root problem.

2. A Proposal of Sibert, Dehornoy and Girault

To get an idea of how a cryptographic system might be founded on CSP and REP (and also to motivate the parameters used in our experiments in Section 3.6), we reproduce an authentication system described in [**SDG02, Sib03**].

2.1. System Description.
The authors of [**SDG02**] refer to the system discussed here as *Scheme III*. It follows the Fiat-Shamir paradigm (cf. [**FS87**]) and involves a *prover* P who chooses a braid $s \in B_n$ along with an integer $m \in \mathbb{N}$. Then P publishes both $b := s^m$ and m, but keeps s secret. During an authentication phase, P wants to convince a *verifier* V that he knows s.

The idea is to convince the verifier V not in the trivial way (namely by sending her s), but rather in *zero-knowledge* (see [**GMR89**]), i.e., without revealing any "usable" information about s. This makes the scheme useful for authentication in the sense that P is the only one who can convince another party of knowing s: even after having convinced V, the braid s remains the prover P's individual secret. For P and V being honest, the communication in Scheme III reads as follows.

(1) P chooses a random $r \in B_n$ and sends $x = rbr^{-1}$ to V.
(2) V chooses a random $\epsilon \in \{0, 1\}$ and sends it to P.
(3) • If $\epsilon = 0$, P sends $y = r$ to V; V then checks $x \stackrel{?}{=} yby^{-1}$,

- if $\epsilon = 1$, P sends $y = rsr^{-1}$ to V; V then checks $x \stackrel{?}{=} y^m$.

The verifier V accepts the proof exactly if the check in the third step succeeds. Of course, extracting an mth root of b will enable anybody to authenticate as P. Also, when communicating with P, choosing $\epsilon = 1$ combined with the ability to solve the CSP (applied to $u = b$ and $w = y^m = rbr^{-1}$) enables an evil verifier to get r and therewith P's indivual secret s. On the other hand, under the assumption that REP and CSP are hard, and additionally some assumptions concerning the choice of instances, [**SDG02**] shows that the scheme indeed is zero-knowledge. That means that no computationally usable information about s is leaked.

2.2. Suggested Parameters. In [**SDG02**], the following parameters are suggested: $n = 30$, and random braids should be chosen either as products of 15 random permutations, or as products of at least 1000 random (Artin) generators. Additionally, to protect against a heuristic approach to the CSP (cf. [**HS02**]), [**Sib03**] suggests to employ cycling and decycling steps in the generation of instances (see [**Sib03**, Algorithme 3.2.2] for details).

3. A Heuristic Apprach to the REP

As outlined in the explanation of the REP, subsequently we assume that $w = u^m \in B_n$ and $m \in \mathbb{N}$ are given, and our goal is to find a $u' \in B_n$ such that $w = u'^m$. Moreover, for most cryptographic applications, infimum and canonical length of u are also either public or at least can be guessed with significant probability. Hence, we will assume $\inf u$ and the canonical length of u to be known, too.

3.1. A First Observation. We start with:

PROPOSITION 3.1. *Say that $u \in B_n$ has Δ-normal form*

$$\Delta^r U_1 \cdots U_s,$$

and that $w := u^m$ has Δ-normal form

$$\Delta^{r'} W_1 \cdots W_{s'}$$

such that $\ell := s' - (m-1)s > 0$. Then the last ℓ canonical factors $W_{(m-1)s+1} \cdots W_{s'}$ of the Δ-normal form of w form a tail of $U_1 \cdots U_s$, i.e.,

$$U_1 \cdots U_s = u' W_{(m-1)s+1} \cdots W_{s'}$$

for some positive $u' \in B_n^+$.

PROOF. Juxtaposing m copies of u and moving powers of Δ to the left, we can write w as

$$(2) \qquad w = \Delta^{mr} W_1^{(1)} \cdots W_{ms}^{(1)},$$

such that for $j = 0, \ldots, m-1$ and $k = 1, \ldots, s$ we have $W_{js+k}^{(1)} = \tau^{r(m-1-j)}(U_k)$, where τ denotes conjugation by Δ. As conjugation by Δ maps canonical factors to canonical factors, all $W_i^{(1)}$ are canonical factors, but the decomposition (2) is of course not necessarily the Δ-normal form of w.

The Δ-normal form of w can then be computed from (2) by successively making decompositions $W_i^{(1)} W_{i+1}^{(1)}$ left-weighted, e.g., using the normal form algorithm in [**CKL**$^+$**01**]. During this process, a number of decompositions

$$(3) \qquad w = \Delta^{r_l} W_1^{(l)} \cdots W_{s_l}^{(l)} \qquad (l > 1)$$

are formulated such that the lth decomposition of w is derived from the $(l-1)^{\text{st}}$ by making exactly one product $W_i^{(l-1)}W_{i+1}^{(l-1)}$ of canonical factors left-weighted. Denote by LR (with L, R canonical factors) the new left-weighted decomposition of $W_i^{(l-1)}W_{i+1}^{(l-1)}$. If $L = \Delta$, then L can be moved to the left (by conjugating all canonical factors $W_j^{(l-1)}$ for $1 \leq j < i$ with Δ and incrementing r_{l-1}). If that happens, or if $R = \varepsilon$, the new decomposition (3) gets shorter in the sense that $s_l < s_{l-1}$.

For some $l = l_0$, this procedure eventually terminates and for that l, the decomposition (3) is the Δ-normal form of w. We claim that for all l with $1 \leq l \leq l_0$, it holds that

$$\text{(4)} \qquad W_{(m-1)s+1}^{(l)} \cdots W_{s_l}^{(l)} \text{ is a tail of } U_1 \ldots U_s.$$

(Note here that by the assumption $s_{l_0} = s' > (m-1)s$ on the canonical length of w and the fact that $s_l \leq s_{l-1}$ for all l, we are guaranteed $s_l \geq (m-1)s + 1$.)

Property (4) can be proven for all l by induction:

- For $l = 1$, statement (4) is clear by construction of $W_{(m-1)s+k}^{(1)} = U_k$ ($1 \leq k \leq s$).
- Assume (4) holds for $l - 1 \geq 1$. Only making a product $W_i^{l-1}W_{i+1}^{l-1}$ left-weighted obviously does not affect property (4). (Here, in the only interesting case $i = (m-1)s$, the very definition of left-weightedness guarantees that (4) still holds.)

  However, by making $W_i^{l-1}W_{i+1}^{l-1}$ left-weighted, the new decomposition LR might collapse in the sense that $L = \Delta$ or $R = \varepsilon$. Then the overall length decreases (i.e., $s_l < s_{l-1}$), canonical factors $W_j^{(l-1)}$ ($i < j \leq s_{l-1}$) get shifted left, and possibly a Δ is "moved through" $W_j^{(l-1)}$ ($1 \leq j < i$).

  For $i \leq (m-1)s$, this does not affect (4), as with $W_{(m-1)s+1}^{(l-1)} \cdots W_{s_{l-1}}^{(l-1)}$ also $W_{(m-1)s+1}^{(l)} \cdots W_{s_l}^{(l)} = W_{(m-1)s+1+k}^{(l-1)} \cdots W_{s_{l-1}}^{(l-1)}$ is a tail of $U_1 \cdots U_s$ (for suitable $k \in \{1, 2\}$). But for $i > (m-1)s$, the case $L = \Delta$ is not possible (because then it would follow that $\Lambda \geq U_1 \cdots U_s$, which contradicts the maximality of $r = \inf u$), and the case $R = \varepsilon$ does not affect (4). Taking it all together, (4) thus holds for l, too.

As just shown, (4) holds in particular for $l = l_0$, and it follows that

$$W_{(m-1)s+1}^{(l_0)} \cdots W_{s_l}^{(l_0)} = W_{(m-1)s+1} \cdots W_{s'}$$

is a tail of $U_1 \ldots U_s$. $\square$

In a nutshell, Proposition 3.1 shows that, given only $w = u^m$, the exponent $m \in \mathbb{N}$ and the canonical length s of u, one can derive a tail of the "canonical part" of u by simply "reading off" canonical factors from the end of the Δ-normal form of w.

3.2. Uncovering the Remaining Part.
Assume we have, e.g., using Proposition 3.1, obtained a tail u_R of the "canonical part" $U_1 \cdots U_s$ of u. Then, let $u_L \in B_n^+$ be the remaining canonical part of u, such that

$$\text{(5)} \qquad u = \Delta^r u_L u_R$$

with $r = \inf u$ as before assumed public. Suppose further that u_L itself is "only" a canonical factor. This latter assumption might seem a little odd, but in our experiments, this turns out to be actually the case for suggested instances of the REP. Note that this implies $\pi^{-1}(\pi(u_L)) = u_L$ for the canonical homomorphism π from Section 1.2.

Consider the projection

$$\overline{u} = \overline{\Delta}^r \, \overline{u_L} \, \overline{u_R} \tag{6}$$

of (5) into the symmetric group S_n, where we write $\overline{x}$ for $\pi(x)$. Obviously, $\overline{u_L}$, and thus by assumption about u_L also $u_L = \pi^{-1}(\overline{u_L})$, is uniquely determined by $\overline{u}$, $\overline{u_R}$ and r. Since r and u_R (and therefore also $\overline{u_R}$) are known, we only need to find $\overline{u}$ to obtain u_L and hence, using (5), also u (which is our final goal).

We can get $\overline{u}$ by solving the equation

$$\overline{w} = x^m \tag{7}$$

in the symmetric group S_n for unknown x. (Note that w and thus $\overline{w}$ is known.) Unfortunately, a solution to this root extraction problem in the symmetric group need not be unique, so there may be many candidates x for $\overline{u}$. However, for every such candidate, (5) can be checked, where $u_L = \pi^{-1}(\overline{u_L})$ is obtained in turn from (6) by setting $\overline{u} = x$. Of course, any solution to (5) solves already the REP in the braid group. Note also that the "right" $\overline{u}$ is among the candidates x for $\overline{u}$, so eventually a solution will be found.

3.3. Extracting Roots of Permutations.

It remains to show how to get *all* solutions $x \in S_n$ to (7) for given $y := \overline{w} \in S_n$ and $m \in \mathbb{N}$. Suppose x consists of disjoint cycles $C_1, \ldots, C_\mu$. Then

$$y = x^m = C_1^m \cdots C_\mu^m,$$

where C_i^m consists of $\gcd(|C_i|, m)$ cycles of length $|C_i|/\gcd(|C_i|, m)$ each.

This allows to treat cycles of y with different lengths separately: let $D_1, \ldots, D_\nu$ be the disjoint cycles of y, and let $\mathcal{L} := \bigcup_{i=1}^\nu \{|D_i|\}$ be the set of cycle lengths occurring in y. Then, any m^{th} root x of y can be expressed as

$$x = \prod_{\ell \in \mathcal{L}} x_\ell, \tag{8}$$

where x_ℓ is an mth root of the product

$$y_\ell := \prod_{|D_i| = \ell} D_i$$

of all cycles of y of length ℓ. Conversely, any product as in (8) with $x_\ell^m = y_\ell$ for all ℓ satisfies $x^m = y$. So without loss of generality, we may assume that y contains only cycles of length ℓ, i.e., that $y = y_\ell$ for some ℓ.

An mth root x of y can then be constructed as follows: choose a divisor $a \geq 1$ of m that is coprime to ℓ such that $g := m/a \leq \nu$. Then $g = \gcd(g\ell, m)$, and thus any g cycles $D_{i_1}, \ldots, D_{i_g}$ can be combined into one larger cycle C of length $|C| = g\ell$ such that $C^m = D_{i_1} \cdots D_{i_g}$. This combination can be done (in ℓ^{g-1} different ways) by writing the D_{i_j} suitably "interleaved". (We omit details here.) The process can be repeated until there are no more cycles D_i left. That way, using backtracking, all possible solutions x can be obtained (however, the number of solutions may of course be large).

3.4. Summary of the Algorithm. In summary, our algorithm thus works as depicted in Figure 1. Some comments are in place: the algorithm takes additionally $\inf u$ and $\sup u$ as input. When not already knowing these values (e.g., because they might be fixed in a cryptographic system), they can simply be guessed: for example, $\inf u$ lies between 0 and $\inf w/m$, and the canonical length s of u lies between 0 and the canonical length s' of w. The algorithm can also be run in parallel for all possible guesses.

Steps 1 and 2 of the algorithm make use of Theorem 3.1 to read off a tail u_R of the canonical factors of u from the Δ-normal form of w. Steps 3 and 4 assume that the remaining canonical part u_L of u is only a canonical factor; then, the method described in Section 3.2 is used to reduce the remaining problem to a root extraction problem in the symmetric group S_n. Step 3 can be implemented by, e.g., using the method described in Section 3.3. Note that a potential solution u' to the REP can easily be checked for correctness, and thus the algorithm never generates wrong output (but it may well output "`fail`" to indicate that something went wrong).

Input: $w = u^m \in B_n$, $m \in \mathbb{N}$, $r = \inf u$, canonical length s of u
Output: either $u' \in B_n$ with $u'^m = w$ or "`fail`"
(1) Let $\ell := \max\{0, s' - (m-1)s\}$ for the canonical length s' of w.
(2) Let $u_R \in B_n$ be the product of the last ℓ canonical factors of w's Δ-normal form.
(3) Solve $\pi(w) = x^m$ in the symmetric group S_n for unknown $x \in S_n$; let $\mathcal{X}$ be the set of solutions.
(4) For all $x \in \mathcal{X}$ do
 • let $u_L := \pi^{-1}(\pi(\Delta)^{-r}\, x\, \pi(u_R)^{-1}) \in B_n$
 • let $u' := \Delta^r u_L u_R$
 • if $u'^m = w$, terminate with output w
end for.
(5) Terminate with "`fail`."

FIGURE 1. Summary of our algorithm

3.5. Improvements and Optimizations. If the algorithm fails, not all is lost. For example, there is a way to make the crucial assumption that the "remaining canonical part" u_L of u is a canonical factor potentially more likely. To see how this works, first observe that to find an mth root of $w \in B_n$, it suffices to find an mth root u' of $w' := \alpha^{-1} w \alpha$, where $\alpha \in B_n$ is arbitrary but known. Namely, then $u := \alpha u' \alpha^{-1}$ is an mth root of w as desired, because

$$u^m = (\alpha u' \alpha^{-1})^m = \alpha u'^m \alpha^{-1} = \alpha \alpha^{-1} w \alpha^{-1} \alpha = w.$$

In other words, we may as well preprocess w by conjugation as long as we keep track of the conjugating braid α. Then it might be useful to *cycle* or *decycle* w (and thus u) so that both end up in their *super summit set* (*SSS*). (Here, cycling and decycling are simply special types of conjugation operations, and the super summit set of a braid w is the subset of w's conjugacy class of braids with minimal canonical length; cf., e.g., [**EM94**] for details. An upper bound on the number of needed cycling/decycling operations is provided in [**BKL01**].) The intuition here is that the SSS contains exactly the "dense" conjugates of u resp. w. One may hope that for these conjugates (call them u' and w') the canonical length of $w' = u'^m$ is close to m times the canonical length of u'. Given that, one can hopefully read off larger tails of u' from the normal form of w' in Step 2 of the algorithm. Also, this makes $\inf u = \inf w/m$ and $\sup u = \sup w/m$ very likely, so that the algorithm can guess r and s with high probability on its own.

Another rather trivial optimization is to not only consider the Δ-normal form (or, left-normal form), but in addition also the right-normal form of w. An obvious analogue to Theorem 3.1 shows then how to read off a head $u_{L'}$ of the canonical part (in the right-normal form) of u from the right-normal form of w. Then it can be hoped that the remaining canonical part $u_{R'}$ of u is only a canonical factor and the problem can be projected into the symmetric group S_n just as in Section 3.2.

3.6. Experimental Results. We have implemented our algorithm (taking into account the optimizations from Section 3.5) in the computer algebra system MAGMA [**BCP97**] on a standard PC. Several different parameter choices have been tested, and in particular we tried to cover concrete suggestions made in the literature (cf. Section 2.2). In Table 1, we have summarized our success rates for extracting *square* roots for different choices of the braid index n and types of instance generation. The table considers the following types of instances:

Artin: Here the MAGMA command `u:=Random(B,r,r)` is used to choose u as a product of r randomly chosen Artin generators (either positive or negative), and our choices for r and n follow [**SDG02**, Section 4.4].

Canonical: Here the MAGMA command `u:=Random(B,0,0,r,r)` is used to choose u as a product of r randomly chosen canonical factors. The parameters $n = 30$, $r = 15$ are taken from [**SDG02**, Section 4.4].

[**Sib03**]: Here u, is chosen according to [**Sib03**, Algorithme 3.2.2] with $\ell = r$. Not being aware of a more concrete suggestion, we adopted the parameter choices from the "Canonical" instances.

n	r	Instance Type	Total samples	Success rate
30	15	Canonical	1000	99.5%
30	15	[**Sib03**]	1000	99.6%
30	1000	Artin	1000	96.7%
60	30	Canonical	500	93.6%
60	30	[**Sib03**]	500	94.0%
60	2000	Artin	500	41.0%

TABLE 1. Experimental results.

It should be noted that "success" indicates that the algorithm found a complete square root of w on input u only[1]. The reason for failure was—independently of parameter choice—almost always the fact that there were too many solutions to (7) to handle them efficiently. In other words, the problem was not the computation in the braid group B_n, but the task of extracting roots in the S_n. So our algorithm seems to succeed almost always in reducing a REP in the B_n to a REP in the S_n. However, note that even to find only one solution to a REP in the B_n, we may have to find *all* solutions to a REP in the S_n.

---

[1] $\inf u$ and $\sup u$ were—after (de)cycling w—guessed as $\inf u = \inf w/2$ and $\sup u = \sup w/2$

This also means that the success rate of the algorithm can be raised significantly (at least for $n > 30$) if it is allowed to run longer (and thus to consider more solutions to (7)). For the computations documented in the above table, the algorithm typically succeeded within seconds.

The results seem to generalize to exponents $m > 2$ and larger braid index n, only the number of solutions to (7) grows substantially with the number of divisors of m. So for large m with many divisors, the attack becomes impractical because not all solutions to (7) can be found efficiently. It should be stressed however, that also here the only problem is the number of solutions in the symmetric group; the reduction to the REP in the S_n itself experimentally works almost always.

3.7. Countermeasures. The heuristic algorithm from [**HS02**] for the conjugacy search problem $w = \alpha^{-1}u\alpha$ in braid groups is closely related to our attack. For their attack, it is crucial that the "distance" of canonically obtained representatives of u and w in their super summit set is small. Actually, the countermeasure from [**Sib03**] against that attack consists in generating u, w with large distance in their SSS. However, this approach of [**Sib03**] does not immediately carry over to our setting, since the braids u, w, u_R, and u_L are not conjugated in general. Additionally, they cannot be chosen separately: in fact, w, u_R, and u_L directly depend on u.

Similarly, choosing braids with a large *ultra summit set* (USS, cf. [**Geb03**]) can prove useful to thwart the algorithm of [**Geb03**] for the conjugacy search problem in braid groups. Again, this approach does not directly relate to our setting, since the braids u and u_R (whose "distance" u_L directly influences whether or not our algorithm succeeds) are neither conjugated in general, nor can they both be chosen separately.

Another, more obvious way to defeat the attack is to provoke a situation in which there is a large number of solutions to the REP (7) in the S_n, so that $\pi(u)$ cannot be found efficiently. As just mentioned, this can be done by choosing large m with a large number of divisors. However, this makes, e.g., the authentication system of [**SDG02**] rather inefficient, because the public key then becomes huge.

Note that choosing, e.g., *pure* braids u (i.e., braids u with $\pi(u) = \mathrm{id}$) in itself does not defeat the attack, as then $\pi(u)$, which is the whole goal of solving (7), is public. An ad hoc strategy which for $m = 2$ successfully counters our attack is the following: Select a pure braid u and insert at random positions in u distinct Artin generators such that $\pi(u)$ consists only of cycles of length 2. Then, $\pi(w) = \pi(u^2) = \mathrm{id}$, so the number of solutions to the REP in S_n is large, and guessing $\pi(u)$ becomes more difficult.

While this choice of instances defeats our attack, we do not endorse the security of this construction and think further investigation of the choice of parameters for the REP in B_n is required. (In particular, in face of our heuristic reduction, what role does the canonical length of u play?)

4. Conclusion

We have described a heuristic algorithm for the root problem in braid groups. This algorithm does not solve the root problem in the general case, yet it applies to most of the cases considered for cryptographic purposes. We have run various experiments with parameters proposed for braid group based cryptosystems to back this result.

Furthermore, we believe that our algorithm can be improved to succeed for some parameters not considered yet for cryptographic applications, and it seems an interesting question how to efficiently find cryptographically satisfying instances of the root problem in braid groups.

Acknowledgments

We thank Markus Grassl for valuable discussions and his help with MAGMA and Robbert de Haan for valuable comments and discussions.

References

[Art25] Emil Artin, *Theorie der Zöpfe*, Abhandlungen aus dem Mathematischen Seminar der Universität Hamburg **4** (1925), 47–72.

[BCP97] Wieb Bosma, John Cannon, and Catherine Playoust, *The Magma algebra system I: The user language*, Journal of Symbolic Computation **24** (1997), 235–265.

[BKL01] Joan S. Birman, Ki Hyoung Ko, and Sang Jin Lee, *The infimum, supremum, and geodesic length of a braid conjugacy class*, Advances in Mathematics **164** (2001), 41–56, Online available at http://mail.konkuk.ac.kr/~sangjin/webfiles/articles/geolen.pdf.

[CJ03] Jung Hee Cheon and Byungheup Jun, *A polynomial time algorithm for the braid Diffie-Hellman conjugacy problem*, Advances in Cryptology, Proceedings of CRYPTO 2003 (Dan Boneh, ed.), Lecture Notes in Computer Science, no. 2729, Springer-Verlag, 2003, Online available at http://eprint.iacr.org/2003/019.ps, pp. 212–225.

[CKL+01] Jae Choon Cha, Ki Hyoung Ko, Sang Jin Lee, Jae Woo Han, and Jung Hee Cheon, *An efficient implementation of braid groups*, Advances in Cryptology, Proceedings of ASIACRYPT 2001 (Colin Boyd, ed.), Lecture Notes in Computer Science, no. 2248, Springer-Verlag, 2001, Online available at http://crypt.kaist.ac.kr/pre_papers/braid-impl.ps, pp. 144–156.

[Deh04] Patrick Dehornoy, *Braid-based cryptography*, Group Theory, Statistics, and Cryptography (Alexei G. Myasnikov, ed.), Contemporary Mathematics, no. 360, American Mathematical Society, 2004, Online available at http://www.math.unicaen.fr/~dehornoy/Surveys/Dgw.ps, pp. 5–33.

[EM94] Elsayed A. Elrifai and H. R. Morton, *Algorithms for positive braids*, Quarterly Journal of Mathematics **45** (1994), 479–497, Online available at http://www.liv.ac.uk/~su14/papers/EM04.ps.gz.

[FS87] Amos Fiat and Adi Shamir, *How to prove yourself: Practical solutions to identification and signature problems*, Advances in Cryptology, Proceedings of CRYPTO '86 (Andrew M. Odlyzko, ed.), Lecture Notes in Computer Science, no. 263, Springer-Verlag, 1987, pp. 186–194.

[Gar69] Frank A. Garside, *The braid group and other groups*, Quarterly Journal of Mathematics **20** (1969), 235–254.

[Geb03] Volker Gebhardt, *A new approach to the conjugacy problem in Garside groups*, Journal of Algebra (2003), To be published, online available at http://arxiv.org/ps/math.GT/0306199.

[GKT+05] David Garber, Shmuel Kaplan, Mina Teicher, Boaz Tsaban, and Uzi Vishne, *Probabilistic solutions of equations in the braid group*, Advances in Applied Mathematics **35** (2005), no. 3, 323–334, Online available at http://arxiv.org/ps/math.GR/0404076.

[GM03] Juan González-Meneses, *The nth root of a braid is unique up to conjugacy*, Algebraic and Geometric Topology **3** (2003), 1103–1118, Online available at http://www.maths.warwick.ac.uk/agt/ftp/main/2003/agt-3-39.ps.

[GMR89] Shafi Goldwasser, Silvio Micali, and Charles Rackoff, *The knowledge complexity of interactive proof systems*, SIAM Journal on Computing **18** (1989), no. 1, 186–208.

[HS02] Dennis Hofheinz and Rainer Steinwandt, *A practical attack on some braid group based cryptographic primitives*, Public Key Cryptography, Proceedings of PKC 2003 (Yvo Desmedt, ed.), Lecture Notes in Computer Science, no. 2567, Springer-Verlag, 2002, pp. 187–198.

[Hug02] Jim Hughes, *A linear algebraic attack on the AAFG1 braid group cryptosystem*, Information Security and Privacy, Proceedings of ACISP 2002 (Lynn Batten and Jennifer Seberry, eds.), Lecture Notes in Computer Science, no. 2384, Springer-Verlag, 2002, Online available at http://www.network.com/hughes/ACISP02.pdf, pp. 176–189.

[KLC+00] Ki Hyoung Ko, Sang Jin Lee, Jung Hee Cheon, Jae Woo Han, Ju-sung Kang, and Choonsik Park, *New public-key cryptosystem using braid groups*, Advances in Cryptology, Proceedings of CRYPTO 2000 (Mihir Bellare, ed.), Lecture Notes in Computer Science, no. 1880, Springer-Verlag, 2000, pp. 166–183.

[LC05] Sunder Lal and Atul Chaturvedi, *Authentication schemes using braid groups*, lanl.arXiv.org ePrint Archive, July 2005, Online available at http://arxiv.org/pdf/cs.CR/0507066.

[Lee04] Sang Jin Lee, *Growth of minimal word-length in Garside groups*, lanl.arXiv.org ePrint Archive, November 2004, Online available at http://arxiv.org/ps/math.GT/0411470.

[SDG02] Hervé Sibert, Patrick Dehornoy, and Marc Girault, *Entity authentication schemes using braid word reduction*, Discrete Applied Mathematics (2002), To be published, online available at http://eprint.iacr.org/2002/187.ps.

[Sib03] Hervé Sibert, *Algorithmique des groupes de tresses*, Ph.D. thesis, Université de Caen, 2003, Online available at http://www.math.unicaen.fr/~sibert/These.pdf.

[Sty79] V.B. Styshnev, *The extraction of a root in a braid group*, Mathematical of the USSR, Izvestija **13** (1979), 405–416.

[SU04] Vladimir Shpilrain and Alexander Ushakov, *The conjugacy search problem in public key cryptography: unnecessary and insufficient*, IACR ePrint Archive, November 2004, Online available at http://eprint.iacr.org/2004/321.pdf.

[Tsa05] Boaz Tsaban, *On an authentication scheme based on the root problem in the braid group*, lanl.arXiv.org ePrint Archive, September 2005, Online available at http://arxiv.org/ps/cs.CR/0509059.

INSTITUT FÜR ALGORITHMEN UND KOGNITIVE SYSTEME, UNIVERSITÄT KARLSRUHE, AM FASANENGARTEN 5, 76131 KARLSRUHE, GERMANY
E-mail address: groch@ira.uka.de

CENTRUM VOOR WISKUNDE EN INFORMATICA, CRYPTOLOGY AND INFORMATION SECURITY GROUP, KRUISLAAN 413, 1098 SJ AMSTERDAM, THE NETHERLANDS
E-mail address: Dennis.Hofheinz@cwi.nl

DEPARTMENT OF MATHEMATICAL SCIENCES, FLORIDA ATLANTIC UNIVERSITY, 777 GLADES ROAD, BOCA RATON, FL 33431, USA
E-mail address: rsteinwa@fau.edu

An attack on a group-based cryptographic scheme

Dennis Hofheinz and Dominique Unruh

ABSTRACT. We give an attack on a public key encryption scheme suggested by Shpilrain and Zapata. Experimental evidence shows that this attack is practical and works for the proposed parameters. We give a way to repair the encryption scheme so that our attack does not work anymore. However, we also expose weak points of the scheme that do not seem to be repairable in an obvious manner.

1. Introduction

Within the last years various attempts have been made to derive cryptographic primitives from problems originating in combinatorial group theory (see, e.g., [**Wag84, WM85, GZ91, AAG99, KLC**[+]**00, AAFG01, Shp04**]). As a relatively new approach, [**Shp04, SZ04**] propose a public key cryptosystem based on metabelian groups. They claim that the security of their scheme is based on the subgroup membership problem in the considered metabelian group.

In this contribution, we show that their scheme can be broken by a very efficient heuristic attack that bypasses solving the subgroup membership problem. This attack uses that public key and ciphertext are transmitted as elements of a free group instead of the considered metabelian group. With the original scheme, this was necessary to allow for en- and decryption. We give a fix to this that allows to have at least the ciphertext transmitted as an element of a metabelian group. With this modification, our attack does not work anymore. But since even with our fix, the public key has to be transmitted as an element of a free group, a reduction to the subgroup membership problem in a metabelian group seems not directly possible. Furthermore, we also expose some additional weak points of the scheme that are also present in the fixed version and do not seem to be easily removable. At the moment, this does not constitute a complete break of the repaired scheme in the sense of a successful attack, but indicates that further research might be needed with respect to some parameters and possibly the proposed platform group.

1991 *Mathematics Subject Classification.* Primary 94A60; Secondary 20F36.

Key words and phrases. Public key cryptography, metabelian groups.

Most of this work was done while the first author was with the Institut für Algorithmen und Kognitive Systeme (IAKS), Lehrstuhl Prof. Beth at the Universität Karlsruhe.

Note. This paper refers to the version [**Shp04, SZ04**] of the Shpilrain-Zapata cryptosystem that was presented at the Canadian Mathematical Society Winter Meeting in December 2004. After presentation of our attack on that system at [**Hof05**], however, the preprint [**SZ04**] was updated (see [**SZ06**]) with improvements very similar to our suggestions, so that the original, unmodified version [**SZ04**] of the Shpilrain-Zapata system is no longer available online. Consequently, our attack from Section 3 does not apply to the updated system [**SZ06**]; however, the observation of weak points in Section 5 does.

2. The Shpilrain-Zapata Cryptosystem

Here, we present the public key cryptosystem of [**Shp04**]. To ease the things to come, we do so in a slightly different (but equivalent) form.

2.1. The System.
First, we describe the system on an abstract level, and in the next subsection, we then discuss some parameter suggestions made in [**Shp04, SZ04**].

Let F_{n+m} be a free group with free generators $x_1, \ldots, x_{n+m}$. Let $R \triangleleft F_{n+m}$ be a normal subgroup of F_{n+m} that is invariant under arbitrary F_{n+m}-endomorphisms. Then $\mathcal{F}_{n+m} := F_{n+m}/R$ is called *relatively free*.

So let for fixed $n, m \in \mathbb{N}$ such $F_{n+m}, R, \mathcal{F}_{n+m}$ be given. Computationally, here any element from $\mathcal{F}_{n+m}$ is given by a free representative (i.e., a word in the free generators x_i and their inverses). Furthermore, any endomorphism $\alpha \in \text{End}(\mathcal{F}_{n+m})$ is given by the images of the generators $x_1 R, \ldots, x_{n+m} R$ under α, so in fact α is represented by a vector of $n+m$ elements from F_{n+m}.

In particular, say that two endomorphisms $\alpha, \beta \in \text{End}(\mathcal{F}_{n+m})$ are given in that way by vectors $(\alpha_1, \ldots, \alpha_{n+m}), (\beta_1, \ldots, \beta_{n+m}) \in F_{n+m}^{n+m}$ such that, e.g., α_i is a free representant of $\alpha(x_i R)$. Then it is clear how to implement the composition $\alpha \circ \beta$: simply substitute every occurence of x_j (resp. x_j^{-1}) in every β_i with α_j (resp. α_j^{-1}).

> **Key generation:** Choose $\varphi \in \text{Aut}(\mathcal{F}_{n+m})$ together with its inverse φ^{-1} such that φ^{-1} cannot be efficiently deduced from φ. (How such a φ is chosen depends on the concrete choice of the underlying subgroup, cf. [**SZ06**].) The public key for encryption is $\hat{\varphi} := \pi_n \circ \varphi$, where $\pi_n \in \text{End}(\mathcal{F}_{n+m})$ is the projection onto the first n generators (i.e., $\pi_n(x_i R) = x_i R$ for $1 \leq i \leq n$ and $\pi_n(x_i R) = 1$ for $n < i \leq n+m$). The secret key for decryption is φ^{-1}.
>
> **Encryption:** A plaintext is an endomorphism $w \in \text{End}(\mathcal{F}_{n+m})$ satisfying $w(x_i R) = 1$ for $n < i \leq n+m$ (such that $w \circ \pi_n = w$). Any such w is encrypted as $c := w \circ \hat{\varphi} \in \text{End}(\mathcal{F}_{n+m})$.
>
> **Decryption:** Decrypting of some $c \in \text{End}(\mathcal{F}_{n+m})$ is done by $w' := c \circ \varphi^{-1}$ such that in case of a legitimately generated ciphertext $c = w \circ \hat{\varphi} = w \circ \pi_n \circ \varphi = w \circ \varphi$, it holds $w' = c \circ \varphi^{-1} = w \circ \varphi \circ \varphi^{-1} = w$ for the decrypted plaintext.

There is a fine point concerning decryption: as φ is not necessarily an automorphism of F_{n+m}, and the equation $\varphi \circ \varphi^{-1} = \text{id}$ does not necessarily hold over F_{n+m}, the free representants of original plaintext and decrypted ciphertext may differ (although they represent the same elements in $\mathcal{F}_{n+m}$). Since as discussed we represent $\mathcal{F}_{n+m}$-endomorphisms by free group elements, the decrypted ciphertext must eventually be put into an $\mathcal{F}_{n+m}$-normal form. So then, actually only the normal form of w (but not its free representation) can be transmitted.

As a remark on this, there is no need to be able to interpret any given plaintext message as a suitable normal form of an element from $\mathcal{F}_{n+m}$. It suffices to be able to choose normal forms in a random way and then to encrypt the actual plaintext with that randomness as a one-time-pad. More specifically, one could encrypt bitstrings m with $m \mapsto (c, H(w) \oplus m)$, where $c = w \circ \varphi$ for a random w, H is a hash function that maps vectors of normal forms to bitstrings, and $\oplus$ denotes the bitwise XOR. (Of course, this simple construction results in a highly malleable scheme [**DDN91**], but then again, efficient constructions are known for converting such a scheme into a non-malleable one, see, e.g., [**FO99**].)

2.2. Suggested Parameters. In [**Shp04, SZ04**], the following parameters were suggested for implementing the above system. First, take $n = 8$ and $m = 2$, such that $F_{n+m} = F_{10}$ is the free group of rank 10. Let $R = [[F_{10}, F_{10}], [F_{10}, F_{10}]]$, such that $\mathcal{F}_{n+m} = F_{n+m}/R$ is the free metabelian group of rank 10.

We omit a description of the way $\hat{\varphi}$ is chosen, as this is not important for our attack. However, it is worthwhile to describe the $\mathcal{F}_{n+m}$-normal form employed during decryption. Namely, it is suggested to use the following normal form $\mathrm{NF}(z)$ for a free representant $z \in F_{n+m}$ of an element from $\mathcal{F}_{n+m}$. First, consider z as an element of the group ring $\mathbb{Z}F_{n+m}$. Let $\mathrm{Fox}_i(z)$ denote the partial Fox derivative with respect to x_i.[1] Let $\mathbf{Fox}(z) = (\mathrm{Fox}_i(z))$ be the vector of the $n+m$ partial Fox derivatives of z. Then $\mathrm{NF}(z)$ is simply the component-wise abelianization of $\mathbf{Fox}(z)$, i.e., $\mathrm{NF}(z) = \left(\overline{\mathrm{Fox}_i(z)}\right)$ where $\overline{a}$ denotes the abelianization of a.

3. Cryptanalysis

Let $\mathcal{F}_{n+m}$ be as before, and let $\hat{\varphi} \in \mathrm{Aut}(\mathcal{F}_{n+m})$ be a public key. Our goal is to decipher a given ciphertext $c = w \circ \hat{\varphi} = w \circ \varphi$. We proceed in two steps: first, we derive the abelianized version $\overline{w}$ of w. Second, we give a heuristic algorithm that, using $\overline{w}$, outputs w with high probability.

At the end of the section, we also give experimental evidence that our approach works. We would like to emphasize that our attack works completely over the free group F_{n+m} and in particular does not solve a subgroup membership problem in a metabelian group. This shows that the system of [**Shp04, SZ04**] is not (at least not solely) based on such a subgroup membership problem, in contrast to what is implied by the title of [**SZ04**].

3.1. The Abelian Part. The crucial observation is that since all computations (apart from the postprocessing of the decrypted ciphertext) take place over the free group F_{n+m}, we know the abelianization of all transmitted group elements. (Note that the abelianization of an element of $\mathcal{F}_{n+m}$ is well-defined, since $R \leq [F_{n+m}, F_{n+m}]$.) So we can assume the abelianizations $\overline{c}$ and $\overline{\hat{\varphi}}$ of c and $\hat{\varphi}$ to be known, and are looking for the abelianization $\overline{w}$ of w. Now the abelianization of $\mathcal{F}_{n+m}$ is isomorphic to $\mathbb{Z}^{n+m}$. Hence, any endomorphism $\overline{\phi}$ of the abelianized $\mathcal{F}_{n+m}$ can be seen as an endomorphism of $\mathbb{Z}^{n+m}$, i.e., as an $(n+m) \times (n+m)$ matrix over $\mathbb{Z}$ acting by left-multiplication on $\mathbb{Z}^{n+m}$. Concretely, the columns of

---

[1] The partial Fox derivative with respect to x_i is the map on $\mathbb{Z}F_{n+m}$ defined by the recursion $\mathrm{Fox}_i(ab) = \mathrm{Fox}_i(a) + a\mathrm{Fox}_i(b)$, $\mathrm{Fox}_i(x_i) = 1$ and $\mathrm{Fox}_i(x_j) = 0$ for $i \neq j$.

this matrix are simply the images of the free abelian generators of under $\overline{\phi}$. In the following, we will thus consider the abelianized versions $\overline{c}, \overline{\hat{\varphi}}, \overline{w}$. of the endomorphisms $c, \hat{\varphi}, w$ as $(n+m) \times (n+m)$ matrices over $\mathbb{Z}$. The set of these matrices we write as $\mathbb{Z}^{(n+m) \times (n+m)}$. By $c = w \circ \hat{\varphi}$ it follows that

$$\overline{c} = \overline{w} \cdot \overline{\hat{\varphi}} \tag{3.1}$$

Further note that the abelianization $\overline{\pi}_n \in \mathbb{Z}^{(n+m) \times (n+m)}$ of π_n is the diagonal matrix with 1 on its first n diagonal elements and 0 on the remaining m diagonal elements. Since $\overline{w} = \overline{w} \cdot \overline{\pi}_n$, it follows that the last m columns of $\overline{m}$ are zero. And since $\overline{\hat{\varphi}} = \overline{\pi}_n \cdot \overline{\hat{\varphi}}$, the last m rows of $\overline{\hat{\varphi}}$ are zero, too. So we can w.l.o.g. consider $\overline{w}$ to be in $\mathbb{Z}^{(n+m) \times n}$ and $\overline{\hat{\varphi}}$ in $\mathbb{Z}^{n \times (n+m)}$. Then (3.1) is an overdetermined system of linear equations. Further, since φ is an automorphism, $\operatorname{rank} \overline{\varphi} = n + m$, and thus $\operatorname{rank} \overline{\hat{\varphi}} = \operatorname{rank} \overline{\pi}_n \cdot \overline{\varphi} = n$. So as an element of $\mathbb{Z}^{n \times (n+m)}$ has full rank and (3.1) has a unique solution $\overline{w}$ which is this original plaintext. This unique solution can then efficiently be found using Gaussian elimination (over $\mathbb{Q}$).

So in summary, we can easily obtain the abelianized version $\overline{w}$ of the plaintext from the public key $\hat{\varphi}$ and an encryption c of w alone.

3.2. The Non-Abelian Part. Let's have a closer look at the encryption operation. Encryption consists of computing $w \circ \hat{\varphi}$ (for public $\hat{\varphi}$) simply as a substitution. More specifically, say that w is given as $(w_1, \ldots, w_{n+m}) \in F_{n+m}^{n+m}$, and $\hat{\varphi}$ as $(\hat{\varphi}_1, \ldots, \hat{\varphi}_{n+m}) \in F_{n+m}^{n+m}$.

Then c is computed as $(c_1, \ldots, c_{n+m}) \in F_{n+m}^{n+m}$, where

$$c_i = \hat{\varphi}_i|_{x_j \to w_j} \tag{3.2}$$

(which means that c_i is equal to $\hat{\varphi}_i$, only that every occurence of any x_j is substituted with the corresponding w_j).

Now say that $\hat{\varphi}_1$ starts with x_j. Then (3.2) means that c_1 starts with w_j (modulo cancellations in the free group). A very simple approach might now be to try to "read off" w_j from the head of c_1. The problems are that (a) cancellations might have taken place and "corrupted" parts of w_j, and (b) there is no telling where w_j finishes and the next $w_{j'}$ starts.

We deal with those two problems by searching different $\hat{\varphi}_i$ that all start with the same generator, e.g., x_j. Although not necessarily the case, the greatest matching prefix of the corresponding c_i can be expected to be a prefix of w_j. Also, such potential prefixes can be found by looking at the tails of c_i for which $\hat{\varphi}_i$ ends on x_j^{-1}. Similarly, one can find potential suffixes of some w_j by looking at the tails of c_i where $\hat{\varphi}_i$ ends with x_j, or at the heads of c_i where $\hat{\varphi}_i$ starts with x_j^{-1}.

As soon as a potential prefix w_j^1 and a suffix w_j^2 of some w_j is found, it can be tried to put w_j together completely. Namely, one can try to chop off generators from the tail of w_j^1 and/or the head of w_j^2 until $w_j^1 w_j^2$ has the correct abelianization $\overline{w}$. (Recall that the previous section shows how to acquire $\overline{w}$.)

Then, as soon as a good candidate for w_j is found, (a) any x_j or x_j^{-1} at the head or the tail of an $\hat{\varphi}_i$ can be eliminated, and (b) the corresponding c_i has to be modified accordingly (i.e., has to be multiplied with the candidate w_j^{-1} or w_j). This yields a simplified system of equations of the type (3.2), in which all $x_j^{\pm 1}$ have been eliminated from the heads and tails of the $\hat{\varphi}_i$. The method described can then be iterated.

Of course, this method is heuristic and heavily relies on the assumption that not too many cancellations in the free group take place. The next subsection gives evidence that nonetheless, our method can be used to successfully attack the system.

3.3. Experimental Results. We have implemented the system in C++ on a standard PC, using the parameters from [**Shp04, SZ04**]. Also, we have implemented the attack described above. We tested several thousand instances, and our algorithm broke the system completely (i.e., correct guess for the complete plaintext w) in about 99% of the cases. The time the attack took ranged from under a second to several minutes, largely depending on the size of the generated public key. (In some rare cases, we even had to abort the key generation, since the memory usage was above one gigabyte.)

4. Foiling the Attack

The attack above needs in an essential way knowledge about free representatives of ciphertext and public key. And not only that, it assumes that encryption took place by performing a variable substitution according to (3.2) in the free group. In fact, the original system was specified exactly like this.

A very obvious way of how to break the assumptions needed for applying our attack would be to actually make use of the relations in $\mathcal{F}_{n+m}$. For example, the ciphertext could be "perturbed" by applying metabelian relations. This method can be combined by changing the presentation (i.e., the relations) of $\mathcal{F}_{n+m}$ as described in [**SZ05**, Section 7]. The problem with this is that there is no obvious way of how to do so *concretely* in a manner that is not invertible by an attacker.

Another way to use these relations would be to transmit the ciphertext as a vector of components in an $\mathcal{F}_{n+m}$-normal form. (In a certain sense, this means applying all relations simultaneously.) However, with the normal form described in [**Shp04, SZ04**] (for different purposes, see above), it is not clear how to decrypt the ciphertext in normal form. Namely, for decrypting c, it has to be composed with the secret key φ^{-1}. Basically, this means substituting all generators in φ^{-1} with the respective components (i.e., images on generators) of c. If these components are in a normal form, it must be possible to multiply two elements in such a normal form. It is not clear how to do so with the normal form from [**Shp04, SZ04**].

To this end, one can simply use a normal form that is multiplicative, in the sense that it is (efficiently) possible to multiply two elements in normal form to get the normal form of the product. The following $\mathcal{F}_{n+m}$-normal form is easily seen to be multiplicative:

For a free representant $z \in F_{n+m}$, let $\mathrm{NF}^*(z)$ be the vector of abelianized Fox derivatives of z, together with the abelianization of z itself. As $R \leq [F_{n+m}, F_{n+m}]$, this is still unique for two representants of the same word $\in \mathcal{F}_{n+m}$.

Furthermore, let two normal forms $\mathrm{NF}^*(z^1), \mathrm{NF}^*(z^2)$ of representatives $z^1, z^2 \in F_{n+m}$ be given. Then the normal form $\mathrm{NF}^*(z^1 z^2)$ of the product consists of the abelianized partial Fox derivatives $\overline{\mathrm{Fox}_i(z^1 z^2)}$ of $z^1 z^2$, and of the abelianized product $\overline{z^1 z^2}$. The latter can be trivially obtained from the abelianized $\overline{z^1}$ and $\overline{z^2}$ (which are part of $\mathrm{NF}^*(z^1)$, resp. $\mathrm{NF}^*(z^2)$), and $\overline{\mathrm{Fox}_i(z^1 z^2)}$ can be computed as

$$\overline{\mathrm{Fox}_i(z^1 z^2)} = \overline{\mathrm{Fox}_i(z^1)} + \overline{z^1} \cdot \overline{\mathrm{Fox}_i(z^2)}$$

by the rules of Fox derivatives.

Because of this efficient multiplication, this normal form can be applied to the ciphertext after encryption. In this way, an attacker does not learn a free representative of the encrypted value, and our attack from above will not work. However, for encryption, the public key still has to be in a free representation: the only obvious way to compute $c = w \circ \hat{\varphi}$ seems to be as a substitution as in (3.2). Here, generators x_j in a free representation of the public key $\hat{\varphi}$ are substituted with the corresponding components w_j of the plaintext w. For this, a free representation of $\hat{\varphi}$ must be available.

5. Further potential weaknesses

We have seen in the previous section how to address our attack by hiding the structure of the free representatives of the transmitted elements of $\mathcal{F}_{n+m}$. However, this may not be sufficient. First of all, note that the abelian part of the attack presented in Section 3.1 still works with the repaired scheme of Section 4. (This is so because for recovering the abelianized solution $\overline{w}$ of the plaintext, only the abelianized $\overline{c}$ and $\overline{\hat{\varphi}}$ are needed.) That is, the abelianized plaintext $\overline{w}$ can still be obtained in the repaired scheme; this might give at least partial information about the plaintext w. It does not seem easy to protect against this, since the abelianized $\overline{c}$ and $\overline{\hat{\varphi}}$ are already uniquely determined by $c, \hat{\varphi} \in \mathcal{F}_{n+m}$; the only way might be to "hide" these abelianizations in a normal form.

Moreover, we conducted experiments using a heuristic algorithm for finding a free representative of an element of $\mathcal{F}_{n+m}$ given in normal form. When a random element x of the free group F_{n+m} of length 500 was chosen, converted to normal form, and then converted back to a free element $\tilde{x}$ using the heuristic algorithm, the probability that $x = \tilde{x}$ was approximately 47%.

To demonstrate the significance of this effect, assume as a thought experiment, that the probability for $x = \tilde{x}$ is near 1 even for long x. Then the improvements mentioned in Section 4 do not help against the attack of Section 3, since we can simply take the normal form and convert it back to the (probably) original element.

However, in reality the situation is not as simple. First, even for a length of 500, the probability of $x = \tilde{x}$ is only 47%, so the probability that all transmitted elements are correctly reconstructed gets exponentially small in the number of the transmitted elements. Second, with increasing length, the probability of $x = \tilde{x}$ seems to fall rapidly (only 22% for length 1000 and 6% for length 2000). But the fact that words of length 500 can be perfectly reconstructed with probability 47% indicates that the relations of $\mathcal{F}_{n+m}$ "strike" rarely, i.e., when considering a random element x of F_{n+m} the shortest representative $\tilde{x}$ of $x + R$ is with high probability similar to x (i.e., large subwords of x and $\tilde{x}$ are identical).

This hypothesis is further supported by the fact that the shortest word in R (except the empty word) has length 14 (in comparison to e.g., 4 for the relations of the free abelian group). Since further the approach of Section 3 could probably be made more fault tolerant by more sophisticated techniques,[2] it is possible that such a procedure might break the cryptosystem even if all transmitted elements

---

[2]Such techniques could include (1) eliminating heads or tails only if there are several indications (and not only one) to support this, (2) backtracking from errors, (3) looking at the interior of the free elements for additional hints, and (4) after each step converting the intermediate results back to the normal form and again to free elements to make use of the simplifications introduced by removing heads or tails (since by only dividing by the head or tail elements, we do not remove errors introduced by the relations of $\mathcal{F}_{n+m}$).

are sent using a normal form or disguised by random application of relations. This might also be considered an indication that the proposed platform group $\mathcal{F}_{n+m}$ is "too close" to a free group for cryptographic purposes.

6. Conclusions

We have shown a way to attack the metabelian group based public key cryptosystem due to Shpilrain and Zapata, and we have verified with experiments that our attack works. We have also shown how to prevent our attack, although even then adaptions of the attack often apply. In summary, we believe that further research is necessary regarding the suggested parameters (and possibly the proposed platform group) for the Shpilrain-Zapata cryptosystem.

References

[AAFG01] Iris Anshel, Michael Anshel, Benji Fisher, and Dorian Goldfeld, *New key agreement protocols in braid group cryptography*, Topics in Cryptology, Proceedings of CT-RSA 2001 (David Naccache, ed.), Lecture Notes in Computer Science, no. 2020, Springer-Verlag, 2001, pp. 13–27.

[AAG99] Iris Anshel, Michael Anshel, and Dorian Goldfeld, *An algebraic method for public-key cryptography*, Mathematical Research Letters **6** (1999), 287–291.

[DDN91] Danny Dolev, Cynthia Dwork, and Moni Naor, *Non-malleable cryptography*, Twenty-Third Annual ACM Symposium on Theory of Computing, Proceedings of STOC 1991, ACM Press, 1991, Extended abstract, full version online available at http://www.wisdom.weizmann.ac.il/~naor/PAPERS/nmc.ps, pp. 542–552.

[FO99] Eiichiro Fujisaki and Tatsuaki Okamoto, *How to enhance the security of public-key encryption at minimum cost*, Public Key Cryptography, Proceedings of PKC '99 (Hideki Imai and Yuliang Zheng, eds.), Lecture Notes in Computer Science, no. 1560, Springer-Verlag, 1999, pp. 53–68.

[GZ91] Max Garzon and Yechezkel Zalcstein, *The complexity of Grigorchuk groups with application to cryptography*, Theoretical Computer Science **88** (1991), no. 1, 83–98.

[Hof05] Dennis Hofheinz, *An attack on a group-based cryptographic scheme*, Invited talk at the 2nd Joint Meeting of AMS, DMV, and ÖMV, Mainz, June 2005.

[KLC+00] Ki Hyoung Ko, Sang Jin Lee, Jung Hee Cheon, Jae Woo Han, Ju-sung Kang, and Choonsik Park, *New public key cryptosystem using braid groups*, Advances in Cryptology, Proceedings of CRYPTO 2000 (Mihir Bellare, ed.), Lecture Notes in Computer Science, no. 1880, Springer-Verlag, 2000, pp. 166–183.

[Shp04] Vladimir Shpilrain, *Combinatorial group theory and public key cryptography*, Invited talk at the Canadian Mathematical Society Winter 2004 Meeting, Montreal, December 2004.

[SZ04] Vladimir Shpilrain and Gabriel Zapata, *Using the subgroup membership search problem in public key cryptography*, Unpublished, superseded by **[SZ06]**, December 2004.

[SZ05] _____, *Combinatorial group theory and public key cryptography*, Applicable Algebra in Engineering, Communication and Computing (2005), To be published, online available at http://eprint.iacr.org/2004/242.ps.

[SZ06] _____, *Using the subgroup membership search problem in public key cryptography*, Algebraic Cryptography (Lothar Gerritzen, Dorian Goldfeld, Martin Kreuzer, Gerhard Rosenberger, and Vladimir Shpilrain, eds.), Contemporary Mathematics, American Mathematical Society, 2006, This volume, online available at http://www.sci.ccny.cuny.edu/~shpil/crypmemb.pdf.

[Wag84] Neal R. Wagner, *Searching for public-key cryptosystems*, IEEE Symposium on Security and Privacy, Proceedings of SSP '84, IEEE Computer Society, 1984, pp. 91–98.

Note also that the unmodified algorithm from Section 3 already has some fault tolerance, since it has to deal with elements corrupted by the cancellation of inverses.

[WM85]   Neal R. Wagner and Marianne R. Magyarik, *A public key cryptosystem based on the word problem*, Advances in Cryptology, Proceedings of CRYPTO '84 (G. Robert Blakley and David Chaum, eds.), Lecture Notes in Computer Science, no. 196, Springer-Verlag, 1985, pp. 19–36.

CENTRUM VOOR WISKUNDE EN INFORMATICA (CWI), KRUISLAAN 413, NL-1090 GB AMSTERDAM, THE NETHERLANDS
  E-mail address: Dennis.Hofheinz@cwi.nl

INSTITUT FÜR ALGORITHMEN UND KOGNITIVE SYSTEME (IAKS), UNIVERSITÄT KARLSRUHE, AM FASANENGARTEN 5, 76131 KARLSRUHE, GERMANY
  E-mail address: unruh@ira.uka.de

Algebraic Problems in Symmetric Cryptography
Two Recent Results on Highly Nonlinear Functions

Nils Gregor Leander

Dedicated to Hans Dobbertin who taught me all this.

ABSTRACT. This survey paper presents two new results in the area of highly nonlinear functions. They will serve as examples demonstrating what kind of algebraic computations can play an important role for the development of results in symmetric cryptography.

1. Introduction

In this survey paper we are going to present some recent results on highly nonlinear functions. Hereby the notion of highly nonlinear will be twofold. In the first part of the paper we are interested in almost perfect nonlinear function from $\mathbb{F}_{2^n}$ to itself. This notion of nonlinearity has its origins in the design of block ciphers that resist *differential attacks*. The second part is devoted to highly nonlinear Boolean functions $f : \mathbb{F}_2^n \to \mathbb{F}_2$, where a function is defined to be highly nonlinear if no good approximation by linear (or affine) Boolean functions exist. This notion plays a central role in the resistance of ciphers against so-called *linear attacks*. Both results that we are going to present in this paper have been found recently (see [1, 15] for more details). They serve as good examples how problems in symmetric cryptography interfere with the algebra of finite fields.

Almost Perfect Nonlinear Functions. In the first part we consider functions
$$F : \mathbb{F}_2^n \to \mathbb{F}_2^n.$$

Every such function can be uniquely represented as a multivariate polynomial, i.e. there exist coefficients $\lambda_u \in \mathbb{F}_2^n$ for $u \in \mathbb{F}_2^n$ such that
$$F(x_1, ..., x_n) = \sum_{u \in \mathbb{F}_2^n} \lambda_u \Big(\prod_{i=1}^n x_i^{u_i}\Big).$$

This representation is called the *algebraic normal form* of F.

2000 *Mathematics Subject Classification.* Primary 94A60.

In this paper we will often identify the vector space $\mathbb{F}_2^n$ with the field $\mathbb{F}_{2^n}$ and consider F as a function from $\mathbb{F}_{2^n}$ to $\mathbb{F}_{2^n}$. From this point of view we can represent F uniquely as a univariate polynomial of degree at most $2^n - 1$, i.e. there exist coefficients $\lambda_i \in \mathbb{F}_{2^n}$ for $i \in \{0, \ldots, 2^n - 1\}$ such that

$$F(x) = \sum_{i=0}^{2^n - 1} \lambda_i x^i.$$

For a function F and any non zero element $c \in \mathbb{F}_2^n$ we define the *derivative* of F with respect to c as

$$\Delta_{F,c} : \mathbb{F}_2^n \rightarrow \mathbb{F}_2^n$$
$$\Delta_{F,c}(x) = F(x) + F(x + c)$$

and the *uniformity* $\mathcal{U}$ of F is defined as

$$\mathcal{U} = \max_{c \neq 0, a \in \mathbb{F}_2^n} |\Delta_{F,c}(a)^{-1}|,$$

where

$$\Delta_{F,c}(a)^{-1} = \{x \in \mathbb{F}_2^n \mid F(x) + F(x+c) = a\}$$

denotes the preimage of a with respect to $\Delta_{F,c}$. The uniformity will serve as our measure for linearity in the first part of the paper.

Highly Non-linear Boolean Functions. In the second part of the paper we are dealing with highly nonlinear Boolean functions and in particular with bent functions. A Boolean function is a mapping

$$f : \mathbb{F}_2^n \rightarrow \mathbb{F}_2.$$

The Walsh-Transform of f is defined by

$$a \in \mathbb{F}_2^n \mapsto f^{\mathcal{W}}(a) = \sum_{x \in \mathbb{F}^n} (-1)^{f(x) + \langle a, x \rangle}$$

The values $f^{\mathcal{W}}(a), a \in \mathbb{F}_2^n$ are called the Walsh coefficients of f. As a measure of the linearity we define the linearity of f as

$$\mathrm{Lin}(f) = \max_{a \in \mathbb{F}_2^n} |f^{\mathcal{W}}(a)|.$$

In this sense a function is highly nonlinear if there is no affine function that is a good approximation, i.e. whose values equal the values of the given Boolean function in most of the possible inputs. This linearity is an important design criterium for the design of both stream- and block-ciphers to resist linear attacks.

2. Almost Perfect Nonlinear Functions

Mappings with uniformity 2 are also called *almost perfect nonlinear* functions, or APN for short. It should be noted that the word *almost* is somewhat misleading as in characteristic two no 1-uniform mapping exist. A mapping is 1-uniform if and only if the mapping $\Delta_{F,c}$ is a bijection for any nonzero c, but in characteristic two we have

$$\Delta_{F,c}(x + c) = F(x + c) + F(x + c + c) = F(x + c) + F(x) = \Delta_{F,c}(x)$$

for all $x \in \mathbb{F}_2^n$. The notion of uniformity plays a central role in differential attacks on block ciphers. This notion was introduced by Nyberg [17] and a lot of work has been done on this notion since their introduction.

The idea, in a nutshell, is to trace how the difference of two encrypted messages m and $m + \delta$ propagates through the different rounds in a block cipher. Basically, if an attacker can guess the output differences with high probability, the cipher will be vulnerable to a differential attack. Thus, a designer of a block cipher has to ensure that given any nonzero input difference no fixed output difference occurs with high probability. This is covered exactly by the notion of uniformity. Since in nearly all block ciphers the only nonlinear part are the so called S-boxes, it is particularly important to study the uniformity of those building blocks.

The property of k-uniformity is invariant under composition with affine mappings. More precisely, let $A, B : \mathbb{F}_2^n \to \mathbb{F}_2^n$ be affine mappings, then a function F is k-uniform if and only if the function

$$F' = A \circ F \circ B$$

is k-uniform. Moreover if F is bijective and k-uniform than the inverse function is also k-uniform. This equivalence is called *extended affine* equivalence. There is a more involved notion of equivalence of k-uniform mappings covering the extended affine equivalence as a special case. This equivalence relation was introduced in [**4**] and is often refereed to as CCZ-equivalence. Two mappings $F, G : \mathbb{F}_2^n \to \mathbb{F}_2^n$ are called CCZ-equivalent if the two graphs, that is the sets

$$\{(x, F(x)) \mid x \in \mathbb{F}_2^n\} \text{ and } \{(x, G(x)) \mid x \in \mathbb{F}_2^n\}$$

are linear equivalent when considered as subsets of $\mathbb{F}_2^n \times \mathbb{F}_2^n$. It was proven in [**4**] that if two mappings are CCZ-equivalent they must have the same uniformity.

From now on we will identify the vector space $\mathbb{F}_2^n$ with the field $\mathbb{F}_{2^n}$ and consider mappings $F : \mathbb{F}_{2^n} \to \mathbb{F}_{2^n}$. Until recently, all known constructions of APN functions happened to be equivalent to power functions, i.e to mappings

$$F : \mathbb{F}_{2^n} \to \mathbb{F}_{2^n}$$
$$x \mapsto x^d$$

for some positive exponent d. See the following table for a complete list of all know exponents.

Exponents d	Reference
$d = 2^k + 1$, $\gcd(k, n) = 1$	proof is trivial
$d = 2^{2k} - 2^k + 1$, $\gcd(k, n) = 1$	[**14**]
$d = 2^{4s} + 2^{3s} + 2^{2s} + 2^s - 1$, $n = 5s$	[**7**]
$d = 2^m + 3$, $n = 2m + 1$	[**11**]
$d = 2^{2r} + 2^r - 1$, $4r + 1 = 0 \bmod n$	[**10**]
$d = -1$, n odd	proof is trivial

It was conjectured that indeed *all* APN functions are equivalent to power functions, until a recent paper [**13**] by Edel, Kyureghyan and Pott introduces a quadratic function from $\mathbb{F}_{2^{10}}$ to itself, which is proved to be inequivalent to any power function. In section 2.1 we are going to present a result, that was found shortly after this work (see [**1**]) and introduced an infinite class of quadratic functions on every number of variables n, divisible by 3, but not by 9.

These recent papers suggest that only the tip of the iceberg of all APN mappings is known. From this perspective one might also wonder why the only known APN functions are constructed using the finite field $\mathbb{F}_{2^n}$ and not only the vector space $\mathbb{F}_2^n$, as indeed the property of being APN is described using the vector space structure only. This might be caused by the fact that these power functions where easier to find and to analyze, and not because most APN functions can be described more naturally using the multiplication of the finite field.

Another important open problem is that currently no APN permutation with an even number of variables is known. The situation for arbitrary mappings is completely unclear. However, for power functions and easy argument shows that if this power function is APN than in the odd case it is necessarily bijective and in the even case it maps 3 to 1. An elegant argument to show that any APN power mapping is bijective when n is odd was given by Dobbertin, and is recalled here. Assume on the contrary that the mapping $x \to x^d$ is not bijective, then there exists an element $x \notin \mathbb{F}_2$ such that $x^d = 1$. Dividing this equality by $(x+1)^d$ yields to

$$\left(\frac{x}{x+1}\right)^d + \left(\frac{1}{x+1}\right)^d = 0 = \left(\frac{x^2}{x^2+1}\right)^d + \left(\frac{1}{x^2+1}\right)^d.$$

Due to the assumption that the power function is APN we must either have

$$\frac{x}{x+1} = \frac{x^2}{x^2+1}$$

or

$$\frac{x}{x+1} = \frac{1}{x^2+1}$$

which leads to $x \in \mathbb{F}_4$. By construction we have $x \notin \mathbb{F}_2$ and thus $x \in \mathbb{F}_4 \setminus \mathbb{F}_2$, a contradiction as n is odd.

On the other hand consider a bijective power mapping when n is even. Than in particular the subfield $\mathbb{F}_4 = \{0, 1, \alpha, \alpha+1\}$ is bijectively mapped to itself and noting that $\sum_{x \in F_4} x = 0$ this implies

$$F(0) + F(1) = F(\alpha) + F(\alpha+1)$$

which contradicts the APN property.

The question if APN permutation exist when n is even, is important for practical aspects. For example the S-Box used in the Advanced Encryption Standard, one of the most widely used encryption algorithm, is not an APN function, as would be desirable for the resistance against the differential attack. This S-Box can be seen as the *world record* as it provides the lowest uniformity that is known for a permutation. Clearly, this situation is not satisfying and thus it would be a major breakthrough in symmetric cryptography to be able to either find an APN permutation or to prove that they do not exist. In the next section we will present the infinite family of APN functions from [1], containing functions which are not equivalent to power mappings.

2.1. A new family of APN functions. The following theorem describes a recently found family of APN functions. We will give a proof in a special case only still demonstrating the algebraic principles of the more general proof.

THEOREM 2.1. *Let s and k be positive integers with $\gcd(s, 3k) = 1$ and $t \in \{1, 2\}$, $i = 3 - t$. Furthermore let $d = 2^{ik} + 2^{tk+s} - (2^s + 1)$,*

$$g_1 = \gcd(2^n - 1, d/(2^k - 1)),$$

$$g_2 = \gcd(2^k - 1, d/(2^k - 1)).$$

If $g_1 \neq g_2$ then the function
$$F : \mathbb{F}_{2^{3k}} \to \mathbb{F}_{2^{3k}}$$
$$x \mapsto \alpha^{2^k-1} x^{2^{ik}+2^{tk+s}} + x^{2^s+1}$$
where α is a primitive element in $\mathbb{F}_{2^{3k}}^*$ is almost perfect nonlinear (and is almost bent when n is odd).

PROOF. We give a proof for the special case $t = 1$ and $s = 1$ here, demonstrating the principle ideas and the algebraic techniques involved. A detailed proof can be found in [**1**].

We have to show that for every $c, d \in L$ the equation
$$F(x) + F(x + c) = d$$
has at most 2 solutions. We have
$$f(x) + f(x+c) = \alpha^{2^k-1} \left(x^{2^{2k}+2^{k+1}} + (x+c)^{2^{2k}+2^{k+1}} \right) + x^3 + (x+c)^3$$
$$(2.1) \qquad = \alpha^{2^k-1} c^{2^{2k}+2^{k+1}} \left(\left(\frac{x}{c}\right)^{2^{2k}} + \left(\frac{x}{c}\right)^{2^{k+1}} \right) + c^3 \left(\left(\frac{x}{c}\right)^2 + \left(\frac{x}{c}\right)\right)$$

As this is a linear equation in x it is sufficient to study the kernel. Note furthermore that
$$c^{2^{2k}+2^{k+1}-3} = c^{(2^k-1)(2^k+3)}$$
and to simplify notation we define
$$a := \left(\alpha c^{2^k+3} \right)^{2^k-1}.$$

After replacing x by cx and dividing by c^3 we finally transferred equation (2.1) into
$$\Delta_a(x) := a \left(x^{2^{2k}} + x^{2^{k+1}} \right) + x^2 + x.$$

We have to prove that for all $c \in L$ this equation has at most two zeros, or equivalently that the only solutions are $x = 0$ and $x = 1$.

The following step can be seen as a very basic application of the multivariate method introduced by Dobbertin [**8**]. If we denote $y = x^{2^k}$, $z = y^{2^k}$ and $b = a^{2^k}$, $c = b^{2^k}$ the above equation $\Delta_a(x) = 0$ can be rewritten as
$$(2.2) \qquad a(y + z^2) + x^2 + x = 0.$$

As a is a $2^k - 1$.th power we have
$$abc = 1.$$

Considering not only equation (2.2) but also the conjugated equations, i.e. the equations derived by taking the 2^k.th and 2^{2k}.th power of equation (2.2), we derive the following system of equations
$$a(z + y^2) + (x + x^2) = 0$$
$$b(x + z^2) + (y + y^2) = 0$$
$$\frac{1}{ab}(y + x^2) + (z + z^2) = 0$$

To show that this system of equations has only the solutions $x = 0$ or $x = 1$ it is sufficient to prove that the resultant of these equations has the desired property. If

we compute the resultant twice, eliminating y and z, we get the following polynomial in x:
$$\frac{a^5}{b}P(a)(x^2+x) = 0.$$
where
$$P(a) = a^3b^3 + a^2b^2 + a^2b + a^2 + ab + b.$$
For $P(a) \neq 0$ this is equivalent to $x \in \mathbb{F}_2$ and thus to prove the theorem it is sufficient to show that $P(a)$ does not vanish for elements a fulfilling the equation

(2.3) $$a = \left(\alpha c^{2^k+3}\right)^{2^k-1}.$$

Note that, if a satisfies (2.3), then a is not a (2^k+3)-th power. Indeed, $g_2 = \gcd(2^k-1, 2^k+3)$ is always a divisor of $g_1 = \gcd(2^n-1, 2^k+3)$. And if a fulfilling (2.3) is a (2^k+3)-th power, then α is a (g_1/g_2)-th power. But as (g_1/g_2) is a nontrivial divisor of $2^n - 1$ this contradicts that α is a primitive element.

Consequently we want to show that if $P(a) = 0$ then a is a $2^k + 3$.th power. But for $\alpha \notin \mathbb{F}_2$ the equation $P(a) = 0$ is equivalent to
$$a = \left(\frac{a+1}{c+1}\right)^3 c^3 \left(\frac{b+1}{a+1}\right) a$$
as can bee seen by expansion and using that $c = 1/ab$. Note that the right hand side is a $2^k + 3$.th power of the element $\frac{a+1}{c+1}c$. This concludes the proof. □

In [1] it was shown that in some cases the APN functions described above are not equivalent to power functions. More precisely the following theorem is stated in [1]

THEOREM 2.2. *Let s and k be positive integers such that $k \geq 4$, $s \leq 3k - 1$, $\gcd(k, 3) = \gcd(s, 3k) = 1$, and $i = sk \mod 3$, $t = 2i \mod 3$, $n = 3k$. Then the function $F(x) = x^{2^s+1} + ax^{2^{ik}+2^{tk+s}}$ with $a \in \mathbb{F}_{2^n}^*$ is EA-inequivalent to power functions on $\mathbb{F}_{2^n}$.*

3. Highly Nonlinear Boolean Functions

Due to the well known Parseval Equation, we have for any Boolean function f
$$\sum_{a \in \mathbb{F}_2^n} f^{\mathcal{W}}(a) = 2^{2n}$$
and thus we have the following bounds
$$2^{n/2} \leq \mathrm{Lin}(f) \leq 2^n$$
Clearly functions obtaining the upper bound are affine. Functions obtaining the lower bound can only exist when n is even and are called *bent* functions, and for the rest of this paper we concentrate on these functions obtaining the highest nonlinearity possible.

Bent functions were introduced by Rothaus [18] in 1976. Because of their own sake as interesting combinatorial objects, but also because of their relations to coding theory and applications in cryptography, they have attracted a lot of research, especially in the last ten years.

Bent functions play a very important role in cryptography. In the design of stream-ciphers or for S-Boxes in block-ciphers, there is a strong need for highly nonlinear functions, to make these ciphers resistant against linear attacks. Due to the fact that high nonlinearity is not the only important criterion in this area, bent functions are usually not directly used, but they serve as a starting point for the construction of highly nonlinear functions that also meet other criteria. For example the best known constructions for highly nonlinear balanced functions, introduced by Dobbertin (see [**9**]), are based on normal bent functions.

Bent functions also play an important role in the area of Reed-Muller Codes. The first order Reed-Muller Code consists of all affine functions on $\mathbb{F}_2^n$ and, if n is even, bent functions on $\mathbb{F}_2^n$ can be characterized as the functions having the maximal possible distance to all the code-words in the first order Reed-Muller Code.

Furthermore Kerdock codes are constructed using (quadratic) bent functions. These nonlinear codes achieve parameters that linear codes cannot achieve.

Another area that is closely related (at least in special cases) is difference sets. Given an abelian (multiplicative) group G of order v, a subset $D \subseteq G$ of order k is called a (v, k, λ)-*difference set* in G, if for each non-identity element g in G, the equation

$$g = xy^{-1}$$

has exactly λ solutions (x, y) in D. It is known that, given a non trivial difference set D in $(\mathbb{F}_2^n, +)$, we always have

(1) n is even,
(2) $k = 2^{n-1} + 2^{n/2-1}$, $\lambda = 2^{n-2} + 2^{n/2-1}$ or
(3) $k = 2^{n-1} - 2^{n/2-1}$, $\lambda = 2^{n-2} - 2^{n/2-1}$.

There is a natural one-to-one correspondence between Boolean functions on $\mathbb{F}_2^n$ and subsets of $\mathbb{F}_2^n$. A Boolean function f on $\mathbb{F}_2^n$ can be characterized by its support, i.e. by the set

$$E_f := \{x \in \mathbb{F}_2^n \mid f(x) = 1\}.$$

As bent functions are precisely the Boolean functions having ideal autocorrelation, it is easy to see that this set E_f is a non-trivial difference set in $\mathbb{F}_2^n$ if and only if f is a bent function. Thus, the open question of characterizing all non-trivial difference sets D in $(\mathbb{F}_2^n, +)$ is equivalent to characterizing all bent functions. For further connections see for example [**5, 6**].

One important question to understand bent function in general was if there exist non-normal bent functions. A Boolean functions for which an affine space of dimension $n/2$ exists such that the restriction of f to this space is constant (resp. affine) is called normal (resp. weakly-normal). The notion of normality was introduced for the first time in [**9**]. While for increasing dimension n a counting argument can be used to prove that nearly all Boolean functions are non-normal, the situation for bent functions is different. Most of the well studied families of bent functions are obviously normal and furthermore, unlike for arbitrary Boolean functions, normality has strong consequences for the behavior of bent functions. One of the consequences is, that if a bent function f is constant on an $\frac{n}{2}$-dimensional affine subspace, then f is balanced on each of the other cosets of this affine subspace. In other words, a normal bent function can be understood as a collection of balanced functions and the question if non-normal bent functions exist, is therefore an important question towards a characterization of bent functions in general. Only a few non-normal bent functions are known so far, see [**2, 3, 2**] for details.

In section 3 we are going to present a class of non-normal bent function from [**15**]. Again it will turn out that the algebraic properties of the finite field of characteristic two play a central role.

For this purpose we identify the vector space $\mathbb{F}_2^n$ with the Galois field $L = \mathbb{F}_{2^n}$. As the notion of a Walsh transform refers to a scalar product, it is convenient to choose the isomorphism such that the canonical scalar product $\langle \cdot, \cdot \rangle$ in $\mathbb{F}_2^n$ coincides with the canonical scalar product in L, which is the trace of the product:

$$\langle x, y \rangle = \sum_{i=1}^{n} x_i y_i = \mathrm{tr}(xy), \quad x, y \in L$$

where

$$\mathrm{tr} : L \mapsto L$$
$$\mathrm{tr}(x) = \sum_{i=0}^{n-1} x^{2^i}.$$

Thus the *Walsh transform* of $f : L \to \mathbb{F}_2$ is defined as

$$f^{\mathcal{W}}(c) = \sum_{x \in L} (-1)^{f(x)} \chi_L(cx), \ c \in L,$$

where

$$\chi_L(x) := (-1)^{\mathrm{tr}_L(x)}$$

is the canonical additive character on L.

We will make extensively use of the following well known property of the trace function

$$\mathrm{tr}_L(x) = \mathrm{tr}_L(x^2).$$

Note that bent functions always occur in pairs. In fact, given a bent function $f : \mathbb{F}_{2^n} \to \mathbb{F}_2$, we define the *dual* f^* of f by the equation

$$(-1)^{f^*(a)} 2^{n/2} = f^{\mathcal{W}}(a),$$

i.e. we consider the signs of the Walsh-coefficients of f.

3.1. Another Class of Non-normal Bent functions.
It was proven by Dobbertin and Dillon in [**12**], using the very powerful concept of Hadamard equivalence, that certain Boolean functions constructed via the derivative of the Kasami Power function are bent. In this section we mainly recall the construction of these functions. For a proof of the bent property see Theorem A in [**12**].

Let $L = \mathbb{F}_{2^n}$ be a finite field of characteristic two where $n = 2k$ denotes an even integer. For any integer r coprime to n the Kasami exponent is defined as

$$d = 2^{2r} - 2^r + 1.$$

Furthermore we denote the derivative of the corresponding power function on L that maps $x \to x^d$ as

$$\Delta_r(x) = (x+1)^d + x^d + 1.$$

Let

$$b_r = L \setminus \Delta_r(L)$$

be the complement of the support of Δ_r, and finally the boolean function

$$c_r^\alpha(x) = B_r(\alpha x^{2^r+1}).$$

where we identify the set $b_r \subset L$ with the boolean function B_r whose support is b_r.

It was proven in [**12**] that for the Walsh-transformation of c_r it holds true that
$$\widehat{c_r^\alpha}(y) = \widehat{f^\alpha}(y^{\frac{2^r+1}{3}})$$
where
$$f^\alpha(x) = \operatorname{tr}(\alpha x^3).$$
is the Gold function. As f is bent whenever $\alpha \in L$ is a non-cube (see Theorem 4.1), it follows that for these α the functions c_r^α are bent. The main goal of this paper is to compute an explicit trace representation of the dual of these functions. The main step therefore is to compute a trace representation of the dual of the function f^α.

Note that instead of working with the complement of the set $\Delta_r(L)$ we could also use the set $\Delta_r(L)$ directly, but for compliance with [**12**] we decided to use the complement as well.

4. The Dual of the c_r Bent Function

In this section we briefly recall the well known Gold-type bent function. We recall a proof of the Gold Case which will allow us to derive a trace representation of the dual, see [**16**].

The monomial bent function belonging to the Gold Exponent is probably the best understood bent function. As it is a quadratic bent function, the dual is quadratic again, and in particular is linear equivalent to the function itself. For the purpose of this paper it is important to compute the corresponding linear mapping (in the special case $d = 3$), as it is done in Lemma 4.2.

THEOREM 4.1. *Let $\alpha \in \mathbb{F}_{2^n}$, $r \in \mathbb{N}$ and $d = 2^r + 1$. The function*
$$f : L \to \mathbb{F}_2$$
with
$$f(x) = \operatorname{tr}(\alpha x^d),$$
is bent if and only if
$$\alpha \notin \{x^d \mid x \in \mathbb{F}_{2^n}\}$$

PROOF. "$\Leftarrow$": Assume that α is not a d.th power. We will prove that f is bent by computing the dual of f.
$$\widehat{f}(a) = 2^{-n/2} \sum_{x \in \mathbb{F}_2^n} \chi_L(\alpha x^d + ax)$$
$$= 2^{-n/2} \sum_{x \in \mathbb{F}_2^n} \chi_L(\alpha(x+\gamma)^d + \alpha\gamma^d + \alpha\gamma^{2^r} x + \alpha\gamma x^{2^r} + ax)$$

for any $\gamma \in \mathbb{F}_2^n$. Assume we could choose γ, such that for every $x \in \mathbb{F}_{2^n}$ we have $\operatorname{tr}(\alpha\gamma^{2^r} x + \alpha\gamma x^{2^r} + ax) = 0$. In this case
$$\widehat{f}(a) = 2^{-n/2} \sum_{x \in \mathbb{F}_2^n} \chi_L(\alpha(x+\gamma)^d + \alpha\gamma^d)$$
$$= 2^{-n/2}(-1)^{\operatorname{tr}(\alpha\gamma^d)} \sum_{x \in \mathbb{F}_2^n} \chi_L(\alpha(x+\gamma)^d)$$
$$= 2^{-n/2}(-1)^{\operatorname{tr}(\alpha\gamma^d)} \widehat{f}(0).$$

So in order to prove that f is bent, we have to consider the linear equation

$$\begin{aligned} 0 &= \text{tr}(\alpha\gamma^{2^r} x + \alpha\gamma x^{2^r} + ax) \\ &= \text{tr}(x^{2^r}(\alpha^{2^r}\gamma^{2^{2r}} + \alpha\gamma + a^{2^r})) \end{aligned}$$

This can only be true for all $x \in \mathbb{F}_{2^n}$ if

$$\alpha^{2^r}\gamma^{2^{2r}} + \alpha\gamma + a^{2^r} = 0.$$

In order to be able to choose γ appropriately, we have to prove that the linear mapping

$$H(\gamma) = \alpha^{2^r}\gamma^{2^{2r}} + \alpha\gamma$$

is bijective, i.e. the mapping has a trivial kernel if $\alpha \notin \{x^d \mid x \in \mathbb{F}_{2^n}\}$. For $\gamma \neq 0$ we compute

$$\begin{aligned} H(\gamma) &= 0 \\ \gamma^{2^{2r}-1} &= \alpha^{1-2^r} \\ (\gamma^d)^{2^r-1} &= (\alpha^{-1})^{2^r-1} \end{aligned}$$

but as $\gcd(2^r - 1, d) = 1$ the left-hand side is a d.th power, while the right-hand side is a d.th power iff α is a d.th power. Thus whenever α is not a d.th power the function is bent.

"$\Rightarrow$": On the other hand this immediately implies, that if α is a d.th power, than f is not bent. Otherwise the function would be bent for every $\alpha \in L^*$ which is not possible. $\square$

If f is bent H^{-1} exists and with $\gamma = H^{-1}(a^{2^r})$ we get

$$\widehat{f}(a) = (-1)^{f(H^{-1}(a))}\widehat{f}(0).$$

Next we concentrate on the special case $r = 3$ and n not divisible by 3. In this case we can without loss of generality choose $\alpha \in \mathbb{F}_{2^2} \setminus \mathbb{F}_2$ and explicitly compute the inverse of the linear mapping H.

LEMMA 4.2. *Let $\gcd(n,3) = 1$ and $\alpha \in \mathbb{F}_{2^2} \setminus \mathbb{F}_2$. Then the inverse of the linear mapping*

$$L(x) = \alpha x^4 + x$$

is given by

$$L^{-1}(x) = \alpha^k \sum_{i=0}^{k-1} \alpha^i x^{2^{2i}}$$

PROOF. The proof is straightforward. We have to show that for all $x \in \mathbb{F}_{2^n}$ it holds that

$$L(L^{-1}(x)) = x.$$

We have:
$$\begin{aligned}
L(L^{-1}(x)) &= \alpha L^{-1}(x^4) + L^{-1}(x) \\
&= \alpha^{k+1} \sum_{i=0}^{k-1} \alpha^i x^{2^{2(i+1)}} + \alpha^k \sum_{i=0}^{k-1} \alpha^i x^{2^{2i}} \\
&= \alpha^k \sum_{i=1}^{k-1} (\alpha^i + \alpha^i) x^{2^{2i}} + (\alpha^{k+1} \alpha^{k-1} + \alpha^k \alpha^0) x \\
&= ((\alpha^k)^2 + (\alpha^k)) x \\
&= x,
\end{aligned}$$

where the last identity comes from the fact that 3 does not divide k and thus $\alpha^k \in \mathbb{F}_4 \setminus \mathbb{F}_2$. □

Note that $H(x) = \alpha L(x)$ and so
$$H^{-1}(x) = L(\alpha^2 x) = \alpha^2 L^{-1}(x)$$
where the last identity follows because L is actually $\mathbb{F}_4$ linear.

We are now able to compute the trace representation of
$$g_r(x) = (c_r^\alpha)^*(x) = (f^\alpha)^* \left(x^{\frac{2^r+1}{3}} \right)$$

As the following computations are indeed independent of r we are going to consider the case $r = 1$ only. We denote g_1 simply by g and with the discussion above we get
$$g(x) = \operatorname{tr}(\alpha \left(L^{-1}(\alpha^2 x) \right)^3).$$
Remember that
$$\operatorname{tr}(x) = \operatorname{tr}(x^2),$$
i.e. we can choose a representant of the cyclotomic equivalence class for each exponent. It turns out that the reduced trace representation of these functions has only a few non-zero coefficients. We skip these lengthy but routine computations.

Finally, we have proven the following theorem (which, as mentioned in the introduction, is stated in a less explicit form in [**12**]).

THEOREM 4.3. *Let $n = 2k$, $d = (2^r + 1)/3$, where $\gcd(r,n) = 1$. Furthermore let α be a primitive element in $GF(4)$ and $u = \lfloor \frac{k-1}{2} \rfloor$.*
(1) *If k is odd then*
$$g_r(x) = \operatorname{tr}\left(\left(\sum_{t=0}^{u-1} \alpha^{2t+1+n} (x^d)^{2^{2t+1}+1} \right) + (u + \alpha^k) x^{2^k+1} \right)$$
is bent.
(2) *If k is even then*
$$g_r(x) = \operatorname{tr}\left(\sum_{t=0}^{u} \alpha^{2t+1+n} (x^d)^{2^{2t+1}+1} \right)$$
is bent.

The dual of these functions is the function derived from the derivative of the Kasami power function c_r. □

Using computer algorithms like described in [2] it turns out that, just like for the monomial bent function corresponding to the Kasami exponent (see [2, 3, 2]), at least some of these functions are non-weakly normal.

FACT 4.4. *For $n = 14$ and $r \neq 1$ the corresponding function is non-weakly normal.*

As a consequence of this fact, these bent functions do not belong to the Maiorana-McFarland class of bent functions. Due to degree reasons these functions do not belong to the Partial-Spread class of bent functions neither. Moreover, as a bent function is weakly normal if and only if the dual is weakly normal, the same holds for the functions c_r^α.

References

[1] L. Budaghyan, C. Carlet, P. Felke, and G. Leander, *An infinite class of quadratic apn functions which are not equivalent to power mappings*, Cryptology ePrint Archive, Report 2005/359, 2005, http://eprint.iacr.org/.

[2] A. Canteaut, M. Daum, H. Dobbertin, and G. Leander, *Normal and non normal bent functions*, International Workshop on Coding and Cryptography, 2003, pp. 91–100.

[3] Anne Canteaut, Magnus Daum, Hans Dobbertin, and Gregor Leander, *Finding nonnormal bent functions.*, Discrete Applied Mathematics **154** (2006), no. 2, 202–218.

[4] Claude Carlet, Pascale Charpin, and Victor Zinoviev, *Codes, bent functions and permutations suitable for des-like cryptosystems.*, Des. Codes Cryptography **15** (1998), no. 2, 125–156.

[5] Chris Charnes, Martin Rötteler, and Thomas Beth, *Homogeneous bent functions, invariants, and designs.*, Des. Codes Cryptography **26** (2002), no. 1-3, 139–154.

[6] U. Dempwolff, *Automorphisms and equivalence of bent functions and of difference sets in elementary abelian 2-groups*, Communications in Algebra **34** (2006), no. 3, 300–305.

[7] H. Dobbertin, *Almost perfect nonlinear power functions on $gf(2^n)$: a new case for n divisible by 5*, Finite Fields and Applications FQ5, 2000, pp. 113–121.

[8] ———, *Uniformly representable permutation polynomials*, Sequences and their applications (SETA), 2001, pp. 1–22.

[9] Hans Dobbertin, *Construction of bent functions and balanced boolean functions with high nonlinearity.*, Fast Software Encryption (Bart Preneel, ed.), Lecture Notes in Computer Science, vol. 1008, Springer, 1994, pp. 61–74.

[10] ———, *Almost perfect nonlinear power functions on gf(2n): The niho case.*, Inf. Comput. **151** (1999), no. 1-2, 57–72.

[11] ———, *Almost perfect nonlinear power functions on $gf(2^n)$: The welch case.*, IEEE Transactions on Information Theory **45** (1999), no. 4, 1271–1275.

[12] Hans Dobbertin and John F. Dillon, *New cyclic difference sets with singer parameters*, Finite Fields and Their Applications **10** (2004), no. 3, 342–389.

[13] Y. Edel, G. Kyureghyan, and A. Pott, *A new apn function which is not equivalent to a power mapping*, IEEE Transactions on Information Theory **52** (2006), no. 2, 744–747.

[14] Heeralal Janwa and Richard M. Wilson, *Hyperplane sections of fermat varieties in ^{3} in char.2 and some applications to cyclic codes.*, AAECC (Gérard D. Cohen, Teo Mora, and Oscar Moreno, eds.), Lecture Notes in Computer Science, vol. 673, Springer, 1993, pp. 180–194.

[15] G. Leander, *Another class of non-normal bent functions*, Second International Workshop on Boolean Functions : Cryptography and Applications, 2006.

[16] ———, *Monomial bent functions*, IEEE Transactions on Information Theory **52** (2006), no. 2, 738–743.

[17] Kaisa Nyberg, *Differentially uniform mappings for cryptography.*, EUROCRYPT, 1993, pp. 55–64.

[18] O. S. Rothaus, *On "bent" functions.*, J. Comb. Theory, Ser. A **20** (1976), no. 3, 300–305.

DEPARTMENT OF MATHEMATICS, RUHR-UNIVERSITY BOCHUM, D-44780 BOCHUM, GERMANY
E-mail address: gregor.leander@ruhr-uni-bochum.de

Inverting the Burau and Lawrence-Krammer Representations

Eonkyung Lee

ABSTRACT. The Burau and Lawrence-Krammer representations have been exploited for braid-group-based cryptography. In particular, it is essential, in many cases of them, to recover preimage braids from images of the representations. In this article, we study how to invert these representations, especially in cryptographers' view.

1. Introduction

Since 1999 [**AAG99**], braid groups have been discussed continuously in cryptography society until recently [**MSU05**]. In looking into the history of braid-group-based cryptography, what is met frequently may be the notion of representations [**AA+01, CJ03, HLP03, Hug02, LP03**]. A representation of a group is a homomorphism from the group to a general linear group. It turns problems in braid groups into those in other groups which can be handled more easily. In order to solve some problems in braid-group-based cryptography, the Burau and Lawrence-Krammer representations have been exploited.

[**Kra00**] proposed a representation, called the Lawrence-Krammer representation (denoted ρ_K), which is faithful[1] for B_4. Its faithfulness for all B_n was proven by [**Big01**] topologically, and then by [**Kra02**] algebraically. In order to solve completely an underlying problem of some cryptographic protocol, [**CJ03**] adopted ρ_K. Using ρ_K, the problem can be solved for all parameters in polynomial time with 100% success probability. An important step toward this achievement is to invert the representation. As a natural approach, [**CJ03**] applied the proof for faithfulness in [**Kra02**] in order to obtain an algorithm inverting ρ_K.

Among all the existing representations of B_n, ρ_K is the only one which is faithful for all n. While ρ_K has such a nice property, dealing with (especially inverting) it is computationally heavy. For instance, $\rho_K(x)$ for $x \in B_n$ is a square matrix of order $O(n^2)$. Compared to it, the Burau representation (denoted ρ_B) is efficient to

---

2000 *Mathematics Subject Classification.* 11T71, 14G50, 94A60.

Key words and phrases. Crytpography, braid group, Burau representation, Lawrence-Krammer representation.

The author was supported in part by the Korea Research Foundation Grant funded by the Korean Government (MOEHRD) (KRF-2005-R04-2004-000-10039-0).

[1]A representation is called *faithful* if it is injective.

compute (e.g. $\rho_B(x)$ for $x \in B_n$ is a square matrix of order n.), and so attracted cryptographers earlier. However, images of ρ_B do not contain enough information about its preimages to convert back a solution obtained in $\rho_B(B_n)$ into B_n. This seems to be related to the unfaithfulness of ρ_B.

In the process of solving an underlying problem of some cryptographic protocol, [**Hug02**] proposed a heuristic algorithm inverting ρ_B. Subsequently, [**LP03**] improved this algorithm both in efficiency and in accuracy, and then heuristically analyzed other cryptographic protocols combining these improved algorithms with the idea from [**HLP03**]. All the proposed algorithms inverting ρ_B work efficiently for the parameters suggested for the corresponding protocols with reasonable success probability.

In this paper, we study how to invert the representations. For the Lawrence-Krammer representation, we show an inverting algorithm which is different from [**CJ03**] but has the same complexity and accuracy. This algorithm is simple and exploits some basic property of braids. For the Burau representation, we survey the inverting algorithms proposed in [**Hug02**] and [**LP03**] with some refinements and comments.

2. Preliminaries

2.1. Notations. For a ring R, $\mathcal{M}_n(R)$ denotes the set of all $n \times n$ matrices over R, and $GL_n(R)$ the general linear group which is the set of all $n \times n$ invertible matrices over R. I_n (resp. 0_n) is the $n \times n$ identity (resp. zero) matrix. For $M \in \mathcal{M}_n(R)$ and $i, j \in \{1, \ldots, n\}$, let $_iM$ (resp. M_i) denote row (resp. column) i of M, and $_iM_j$ the entry in row i and column j of M. $R[t_1, \ldots, t_n]$ denotes the ring composed of polynomials in $(t_1, \ldots, t_n)$ with coefficients in R.

2.2. Braid Groups. The n-braid group B_n can be presented by the $n-1$ Artin generators $\sigma_1, \ldots, \sigma_{n-1}$ and two kinds of relations $\sigma_i \sigma_j = \sigma_j \sigma_i$ for $|i-j| > 1$ and $\sigma_i \sigma_j \sigma_i = \sigma_j \sigma_i \sigma_j$ for $|i-j| = 1$.

The submonoid B_n^+ of B_n is the set of all positive words with the identity e_n. For $x \in B_n^+$, $|x|$ denotes the word length of x.

A braid $(\sigma_1 \cdots \sigma_{n-1})(\sigma_1 \cdots \sigma_{n-2}) \cdots (\sigma_1)$ is called the *fundamental braid* of B_n and denoted Δ. The *infimum* of x, denoted $\inf(x)$, is an invariant of x. It indicates the largest integer u such that there exists $Q \in B_n^+$ satisfying $x = \Delta^u Q$.

3. The Lawrence-Krammer Representation

Throughout this section, we let $m = n(n-1)/2$. Let V be a free $\mathbb{Z}[t^{\pm 1}, q^{\pm 1}]$-module with rank m, and let $\{v_{j,k}\}_{1 \le j < k \le n}$ be its basis. Then $GL_m(\mathbb{Z}[t^{\pm 1}, q^{\pm 1}])$ is the set of all automorphisms on V. The Lawrence-Krammer representation $\rho_K \colon B_n \to GL_m(\mathbb{Z}[t^{\pm 1}, q^{\pm 1}])$ is defined by

$$\rho_K(\sigma_i)(v_{j,k}) = \begin{cases} v_{j,k} & \text{if } i \ne j-1, j, k-1, k, \\ (q^2-q)v_{j-1,j} + qv_{j-1,k} + (1-q)v_{j,k} & \text{if } i = j-1, \\ v_{j+1,k} & \text{if } i = j \ne k-1, \\ qv_{j,k-1} + (1-q)v_{j,k} + t(q^2-q)v_{k-1,k} & \text{if } i = k-1 \ne j, \\ v_{j,k+1} & \text{if } i = k, \\ tq^2 v_{j,k} & \text{if } i = j = k-1, \end{cases}$$

Algorithm 1 Inverting ρ_K

Input: $X = \rho_K(x)$ where $x \in B_n$
Output: x
1: $u \leftarrow \mu(X);\ X \leftarrow \rho_K(\Delta^{-u})X;\ x \leftarrow e_n$
2: **while** $X \neq I_m$ **do**
3:   **for** $i = 1$ to $n-1$ **do**
4:     Compute $Y \leftarrow \rho_K(\sigma_i^{-1})X$.
5:     **if** $\mu(Y) = 0$ **then**
6:       $x \leftarrow x\sigma_i;\ X \leftarrow Y;$ break
7:     **end if**
8:   **end for**
9: **end while**
10: $x \leftarrow \Delta^u x$
11: Return(x)

---

for all $i \in \{1, \ldots, n-1\}$ and for all $j, k \in \{1, \ldots, n\}$ such that $j < k$. Images of braids under ρ_K are called the *Lawrence-Krammer matrices*.

ρ_K is faithful for all n. So, there is a unique braid when its Lawrence-Krammer matrix is given. Krammer [**Kra02**] showed that the Laurent expansion of a given Lawrence-Krammer matrix with respect to t provides some information about the preimage braid, from which the following can be deduced.

FACT 3.1. *For $\rho_K(x)$, $\inf(x)$ equals the greatest common degree of t of all the m^2 entries of $\rho_K(x)$. In other words,*

$$\inf(x) = \max\left\{u \in \mathbb{Z} \mid \rho_K(x) = At^u,\ A \in \mathcal{M}_m(\mathbb{Z}[t, q^{\pm 1}])\right\}.$$

3.1. An Inverting Algorithm. The following proposition shows a useful property of infimum, which is almost the same as Proposition 1 in [**LLH01**] (so we omit the proof). We apply it in combination with Fact 3.1 to invert ρ_K.

PROPOSITION 3.2. *Let $x \in B_n^+ - \Delta B_n^+$. Then, for any i*

$$\inf(\sigma_i^{-1}x) = \begin{cases} 0 & \text{if } \sigma_i \text{ is a first generator of } x, \\ -1 & \text{otherwise.} \end{cases}$$

For any $\rho_K(x)$ and any i, we can compute $\rho_K(\sigma_i^{-1}x)$, and so $\inf(\sigma_i^{-1}x)$ by Fact 3.1. So, for any $x \in B_n^+ - \Delta B_n^+$, a first generator of x can be discovered by comparing $\inf(\sigma_i^{-1}x)$ and 0 by the above proposition.

For any $x \in B_n$, $\Delta^{-\inf(x)}x$ is a positive braid with zero infimum (i.e. $\Delta^{-\inf(x)}x \in B_n^+ - \Delta B_n^+$). So, if we can compute x from $\rho_K(x)$ for all $x \in B_n^+ - \Delta B_n^+$, we can do the same task for all $x \in B_n$ since $\inf(x)$ can be obtained from $\rho_K(x)$.

Define a function $\mu : \rho_K(B_n) \to \mathbb{Z}$ by $\mu(\rho_K(x)) = \inf(x)$ for $x \in B_n$. From $\rho_K(x)$, the value of $\mu(\rho_K(x))$ can be computed efficiently. On input $\rho_K(x)$ for an unknown $x \in B_n$, Algorithm 1 reconstructs x by computing $\Delta^{-\inf(x)}x$ from $\rho_K(x)$ generator by generator from left to right. By Proposition 3.2, this algorithm succeeds with 100% probability.

3.2. Complexity. From the definition of $\rho_K(\sigma_i)$, we can see that $\rho_K(\sigma_i)$ is the same as I_m except only $2n-3$ columns. In other words, $m-2n+3$ columns of $\rho_K(\sigma_i)$ are each in the form of $0 \cdots 010 \cdots 0$ where the '1' lies on the main diagonal of $\rho_K(\sigma_i)$. So, the number of non-zeros on each row of $\rho_K(\sigma_i^{-1})$ is at most $2n-2$.

Algorithm 2 Finding x from $\rho_B(x)$

Input: $X \in \rho_B(B_n^+)$
Output: $z \in B_n^+$
 1: **if** $X = I_n$ **then**
 2:    $z \leftarrow e_n$
 3: **else**
 4:    Compute l such that $\det(X) = (-t)^l$.
 5:    **for** i from l to 1 by -1 **do**
 6:      Compute $j_c = \mathsf{C_H}(X)$.
 7:      **if** $j_c = n$ **then**
 8:        break
 9:      **end if**
10:      $A[i] \leftarrow j_c;\ X \leftarrow X\rho_B(\sigma_{j_c})^{-1}$
11:    **end for**
12:    $z \leftarrow \sigma_{A[i+1]} \cdots \sigma_{A[l]}$
13: **end if**
14: Return(z)

What most affects the complexity of Algorithm 1 is the number of multiplications in $\mathbb{Z}[t^{\pm 1}, q^{\pm 1}]$. By the above observation, it equals $O(m^3 + |\Delta^{-\inf(x)}x|nm^2n) = O(|\Delta^{-\inf(x)}x|n^6)$, which is eventually the same as the algorithm presented in Section 3.2 of [**CJ03**].

4. The Burau Representation

The Burau representation $\rho_B : B_n \to GL_n(\mathbb{Z}[t^{\pm 1}])$ is defined by

$$\rho_B(\sigma_i) = \mathrm{diag}\left(I_{i-1}, \begin{bmatrix} 1-t & t \\ 1 & 0 \end{bmatrix}, I_{n-i-1}\right)$$

for all $i = 1, \ldots, n-1$. This representation is known to be unfaithful for all $n \geq 5$. Images of braids under ρ_B are called the *Burau matrices*. For all $i = 1, \ldots, n-1$, $\det(\rho_B(\sigma_i)) = -t$.

When applying braids to cryptography, $\rho_B(x)$ is usually given together with the range of x, and so with the lower bound, say u, of $\inf(x)$. Since $x' = \Delta^{-u}x \in B_n^+$, we can compute x from $\rho_B(x)$ if we can do the same task for x'. So, we may deal with only positive braids for inverting ρ_B.

4.1. Hughes' Algorithm. [**AAG99**] introduced in general terms a key agreement protocol based on combinatorial group theory. Subsequently [**AA+01**] showed how to implement it using braid groups.

In analyzing the security of this scheme, Hughes [**Hug02**] proposed and used an algorithm inverting ρ_B. Note that all the entries of $\rho_B(x)$ are polynomials in t if $x \in B_n^+$. His main idea is to reconstruct x from $\rho_B(x)$ generator by generator from right to left by assuming that the first column with the highest-degree entry in $\rho_B(x)$ indicates a last generator of x. For the parameters suggested in [**AA+01**], it succeeds with 96% out of 40 runs very fast.

We describe his algorithm in Algorithm 2. For $x \in B_n$, let $\mathsf{C_H}(\rho_B(x))$ denote the integer pointing out the first column containing the highest-degree entry in $\rho_B(x)$. If there is no confusion from the context, it is interchangeably written as $\mathsf{C_H}(x)$.

Algorithm 3 Finding x from $\rho_B(x)$ without self-correction

Input: $X \in \rho_B(B_n^+)$
Output: $z \in B_n^+$
1: **if** $X = I_n$ **then**
2:     $z \leftarrow e_n$
3: **else**
4:     Compute l such that $\det(X) = (-t)^l$.
5:     **for** i from l to 1 by -1 **do**
6:         Compute $j_c = \mathsf{C}_{\mathsf{LP}}(X)$.
7:         **if** there does not exist such j_c **then**
8:             break
9:         **else**
10:            $A[i] \leftarrow j_c$; $X \leftarrow X\rho_B(\sigma_{j_c})^{-1}$
11:         **end if**
12:     **end for**
13:     $z \leftarrow \sigma_{A[i+1]} \cdots \sigma_{A[l]}$
14: **end if**
15: Return(z)

In Algorithm 2, it is checked whether or not $\mathsf{C}_\mathsf{H}(X) = n$ since $\mathsf{C}_\mathsf{H}(X)$ points out an invalid generator if $\mathsf{C}_\mathsf{H}(X) = n$, as [**Hug02**] mentioned. (This contains a possibility that we have done well until that point, however, we cannot go on farther at that point.)

As well as this original reason, we note that $\mathsf{C}_\mathsf{H}(X) = n$ is a sign of past failure. Assume that $A[i+1], \ldots, A[l]$ are all guessed correctly for some i. In this case, the remaining part of the braid (i.e. the preimage braid of X in the i-th round) is a positive braid with word length i. From the lemma below, $\mathsf{C}_\mathsf{H}(X)$ in the i-th round cannot be n. As a result, $\mathsf{C}_\mathsf{H}(X) = n$ indicates that we have ever removed a wrong generator before $\mathsf{C}_\mathsf{H}(X) = n$ happens, and so are going the wrong way.

LEMMA 4.1. [**Lee04**, Lemma 1] *For all* $x \in B_n^+$, $\mathsf{C}_\mathsf{H}(x) < n$.

Note that the above lemma does not hold without restriction to positive braids. For instance, $\mathsf{C}_\mathsf{H}(\sigma_1^{-2}\sigma_2^{-1}) = 3$ in B_3.

4.2. Lee-Park's Algorithms. [**KL+00**] proposed a public-key encryption scheme based on braid groups. Subsequently [**CK+01**] generalized it providing computing tool kits. In analyzing the security of these schemes, Lee and Park [**LP03**] explored the task of inverting ρ_B. For Hughes' empirical evidence that $\mathsf{C}_\mathsf{H}(x)$ tends to indicate a last generator of x, they theoretically investigated the reason, and then proposed two algorithms, Algorithm 3 and Algorithm 4, by improving his algorithm. (The two algorithms are described here a little differently from the original versions for refinements.)

Algorithm 3 is faster (in experience) and succeeds with higher probability (in theory, but almost the same in practice) than Algorithm 2. Algorithm 4 succeeds with higher probability but is slower than Algorithm 2. The objective of these algorithms is to seek an unknown sequence $A[1], \ldots, A[|x|]$ satisfying $x = \sigma_{A[1]} \cdots \sigma_{A[|x|]}$ from $\rho_B(x)$.

Tables 1 and 2 show some experimental results for Algorithms 3 and 4 comparing with Algorithm 2. The experiment was performed on a computer with a

Algorithm 4 Finding x from $\rho_B(x)$ with self-correction

Input: $X \in \rho_B(B_n^+)$
Output: $z \in B_n^+$
1: **if** $X = I_n$ **then**
2:    $z \leftarrow e_n$
3: **else**
4:    Compute l such that $\det(X) = (-t)^l$.
5:    $M[l] \leftarrow X$
6:    **for** i from l to 1 by -1 **do**
7:      Compute $j_a = \mathsf{C_H}(M[i])$ and $j_c = \mathsf{C_{LP}}(M[i])$.
8:      **if** there exists such j_c and $j_c = j_a$ **then**
9:        $A[i] \leftarrow j_c$; $M[i-1] \leftarrow M[i]\rho_B(\sigma_{A[i]})^{-1}$
10:      **else**
11:        **if** $i = l$ **then**
12:          break
13:        **end if**
14:        **if** there is k $(> i)$ such that $j_c = j_a > 1$ for $M[k]$, $A[k] = j_c$, and every entry of $M[k]_{j_c}$ is in $t\mathbb{Z}[t]$ **then**
15:          reset i to be the smallest value among such k's;
16:          $A[i] \leftarrow A[i] - 1$; $M[i-1] \leftarrow M[i]\rho_B(\sigma_{A[i]})^{-1}$
17:        **else**
18:          break
19:        **end if**
20:      **end if**
21:    **end for**
22:    **if** $i = l$ **then**
23:      $z \leftarrow e_n$
24:    **else**
25:      $z \leftarrow \sigma_{A[i+1]} \cdots \sigma_{A[l]}$
26:    **end if**
27: **end if**
28: Return(z)

Pentium IV 1.7 GHz processor using Maple 7. On input (n, l), the program chooses at random 500 x's from B_n^+ with $|x| = l$, computes $\rho_B(x)$ from x, computes z from $\rho_B(x)$ by each algorithm, and then checks whether or not z is equal to x by comparing their normal forms. Below we explain Algorithms 3 and 4 in more detail.

Algorithm 3: If σ_j is a last generator of x, then every entry in $\rho_B(x)_{j+1}$ is always in $t\mathbb{Z}[t]$ because we deal with only positive braids here. Let $\mathsf{C_{LP}}(\rho_B(x))$ denote the integer pointing out the first column containing the highest-degree entry in $\rho_B(x)$ among the columns whose next column's entries are all in $t\mathbb{Z}[t]$. Like $\mathsf{C_H}(\cdot)$, let $\mathsf{C_{LP}}(\rho_B(x))$ and $\mathsf{C_{LP}}(x)$ be used interchangeably.

Although Algorithm 3 has an additional task to Algorithm 2 by checking whether or not $_k\rho_B(x)_{j+1} \in t\mathbb{Z}[t]$ for all $k = 1, \ldots, n$ at Step 6, it is eventually more efficient than Algorithm 2 as seen in Table 1 [**LP03**, Table 1]. In experiments, due to this task, Algorithm 3 turns out to stop on the way to building z usually much before $|z| = |x|$ in the case of $z \neq x$. Namely, $i \gg 0$ at Step 13. Meanwhile, Algorithm 2 goes on much farther. As [**Hug02**, pp.182] mentioned, it

TABLE 1. Elapsed time in recovering x from $\rho_B(x)$ (unit: millisecond)

$(n, |x|)$	(7, 40)	(7, 55)	(7, 70)	(10, 60)	(10, 80)	(10, 100)
Algorithm 2	260	368	484	515	745	1060
Algorithm 3	252	324	414	487	651	790

(Note that this table compares only the inverting processes themselves. The time commonly taken in computing $|x|$ from $\rho_B(x)$ is not included.)

goes on until $|z| = |x|$ in experiments. So the time gap between the two algorithms grows as $|x|$ increases.

Algorithm 4: Algorithm 4 is an upgraded version of Algorithm 3 in accuracy. Given a positive braid x, let $x_1 = \sigma_{\mathsf{C_{LP}}(x)}$, $x_2 = \sigma_{\mathsf{C_{LP}}(xx_1^{-1})}$, ..., $x_k = \sigma_{\mathsf{C_{LP}}(xx_1^{-1} \cdots x_{k-1}^{-1})}$ for $k < |x|$, and let $x' = xx_1^{-1} \cdots x_k^{-1}$. Empirically, if $\mathsf{C_{LP}}(x') \neq \mathsf{C_H}(x')$, then $\mathsf{C_{LP}}(x'x_k \cdots x_i)$ does not indicate a last generator of $x'x_k \cdots x_i \overset{\text{let}}{=} y$ for some $1 \leq i \leq k$. The idea of Algorithm 4 is that, for this y, if $\mathsf{C_{LP}}(y) > 1$ and every entry in the $\mathsf{C_{LP}}(y)$-th column of $\rho_B(y)$ is in $t\mathbb{Z}[t]$, then $\mathsf{C_{LP}}(y) - 1$ has a good possibility to indicate a last generator of y. As Table 2 [**LP03**, Table 2] shows, the success probability of Algorithm 4 is higher than Algorithm 3. However, it is slower due to this self-correction process.

An anonymous referee of this article reproduced the experimental results for Algorithm 4 on a better platform using a more optimized software, and reported that the success probabilities of Algorithm 4 are in fact much higher than those in Table 2.

TABLE 2. Success rate of recovering x from $\rho_B(x)$ (unit: %)

n	5			7			10		
$|x|$	30	40	50	40	55	70	60	80	100
Algorithm 3	96	83	76	91	76	64	87	67	42
Algorithm 4	100	99	97	99	97	82	99	90	69

4.3. Comparison of Inverting Algorithms.

From the above results, we have Table 3 that compares Algorithms 2, 3, and 4 with respect to their success rate and elapsed time. The table is valid for all parameter sets not being limited to those in Tables 1 and 2.

TABLE 3. Comparison of algorithms recovering x from $\rho_B(x)$

Success rate	Algorithm 2	$\leq^\dagger$	Algorithm 3	<	Algorithm 4
Elapsed time	Algorithm 3	<	Algorithm 2	<	Algorithm 4

($\dagger$ holds in theory, but Algorithm 2 $\approx$ Algorithm 3 in experiments.)

Acknowledgements

The author would like to thank gratefully an anonymous referee. (S)he pointed out the missing of $A[k] = j_c$ on the line 14 of Algorithm 4, and that Algorithm 4 is much powerful than known before.

References

[AA+01] I. Anshel, M. Anshel, B. Fisher, and D. Goldfeld. *New Key Agreement Protocols in Braid Group Cryptography.* CT-RSA '01, LNCS **2020** (2001), 13–27.

[AAG99] I. Anshel, M. Anshel, and D. Goldfeld. *An Algebraic Method for Public-Key Cryptography.* Math. Res. Lett. **6** (1999), 1–5.

[Big01] S. Bigelow. *Braid Groups are Linear.* J. Amer. Math. Soc. **14** (2001), 471–486.

[CK+01] J.C. Cha, K.H. Ko, S.J. Lee, J.W. Han, and J.H. Cheon. *An Efficient Implementation of Braid Groups.* ASIACRYPT '01, LNCS **2248** (2001), 144–156.

[CJ03] J.H. Cheon and B. Jun. *A Polynomial Time Algorithm for the Braid Diffie-Hellman Conjugacy Problem.* CRYPTO '03, LNCS **2729** (2003), 212–225.

[HLP03] S.G. Hahn, E. Lee, and J.H. Park. *Complexity of the Generalized Conjugacy Problem.* Discrete Applied Mathematics **130** (1) (2003), 33–36.

[Hug02] J. Hughes. *A Linear Algebraic Attack on the AAFG1 Braid Group Cryptosystem.* ACISP '02, LNCS **2384** (2002), 176–189.

[KL+00] K.H. Ko, S.J. Lee, J.H. Cheon, J.W. Han, J.S. Kang, and C. Park. *New Public-Key Cryptosystem Using Braid Groups.* CRYPTO '00, LNCS **1880** (2000), 166–183.

[Kra00] D. Krammer. *The Braid Group B_4 is Linear.* Inventiones Mathematicae **142** (2000), 451–486.

[Kra02] D. Krammer. *Braid Groups are Linear.* Annals Mathematics **155** (2002), 131–156.

[Lee04] E. Lee. *Braid Groups in Cryptology.* IEICE Trans. Fundamentals **E87-A** No. 5 (2004), 986–992.

[LLH01] E. Lee, S.J. Lee, and S.G. Hahn. *Pseudorandomness from Braid Groups.* CRYPTO '01, LNCS **2139** (2001), 486–502.

[LP03] E. Lee and J.H. Park. *Cryptanalysis of the Public-Key Encryption based on Braid Groups.* EUROCRYPT '03, LNCS **2656** (2003), 477–490.

[MSU05] A. Myasnikov, V. Shpilrain, and A. Ushakov. *A Practical Attack on Some Braid Group Based Cryptographic Protocols.* CRYPTO '05, LNCS **3621** (2005), 86–96.

DEPARTMENT OF APPLIED MATHEMATICS, SEJONG UNIVERSITY, SEOUL, 143-747, KOREA
E-mail address: eonkyung@sejong.ac.kr

A new key exchange protocol based on the decomposition problem

Vladimir Shpilrain and Alexander Ushakov

ABSTRACT. In this paper we present a new key establishment protocol based on the decomposition problem in non-commutative groups which is: given two elements w, w_1 of the platform group G and two subgroups $A, B \subseteq G$ (not necessarily distinct), find elements $a \in A$, $b \in B$ such that $w_1 = awb$. Here we introduce two new ideas that improve the security of key establishment protocols based on the decomposition problem. In particular, we conceal (i.e., do not publish explicitly) one of the subgroups A, B, thus introducing an additional computationally hard problem for the adversary, namely finding the centralizer of a given finitely generated subgroup.

1. Introduction

In search of a more efficient and/or secure alternative to established cryptographic protocols (such as RSA), several authors have come up with public key establishment protocols as well as with complete public key cryptosystems based on allegedly hard *search problems* from combinatorial (semi)group theory, including the conjugacy search problem [1, 15], the homomorphism search problem [14], [18], the decomposition search problem [5, 15, 17], the subgroup membership search problem [19].

In this paper, we focus on the decomposition search problem which we subsequently call just the decomposition problem. The problem is: given two elements w, w_1 of the platform group G and two subgroups $A, B \subseteq G$ (not necessarily distinct), find elements $a \in A$, $b \in B$ such that $w_1 = awb$.

It is straightforward to arrange a key establishment protocol based on this problem (see [5, 15, 17]), assuming that $ab = ba$ for any $a \in A$, $b \in B$:

(0) One of the parties (say, Alice) publishes a random element $w \in G$ (the "base" element).

(1) Alice chooses $a_1, a_2 \in A$ (Alice's private keys) and sends $a_1 w a_2$ to Bob.

(2) Bob chooses $b_1, b_2 \in B$ (Bob's private keys) and sends $b_1 w b_2$ to Alice.

(3) Alice computes
$$K_a = a_1 b_1 w b a_2 b_2$$
and Bob computes
$$K_b = b_1 a_1 w a_2 b_2.$$
If $a_i b_i = b_i a_i$, then $K_a = K_b$ in G. Thus Alice and Bob have a *shared secret key*.

2000 *Mathematics Subject Classification.* 94A60, 20F05, 20F36, 68P25.
Research of the first author was partially supported by the NSF grant DMS-0405105.

Security of such a protocol will, of course, depend on a particular platform group G (at the very least, G has to be non-commutative). It appears that for braid groups (which are a popular choice for the platform), the so-called *length attacks* present a serious threat, see e.g. [8, 12, 13, 16].

In this paper, we introduce two new ideas that improve the security of key establishment protocols based on the decomposition problem:

(i) We conceal one of the subgroups A, B.

(ii) We make Alice choose her left private key a_1 from one of the subgroups A, B, and her right private key a_2 from the other subgroup. Same for Bob.

These two improvements together will obviously foil any length attacks. We give a complete description of our protocol in the following Section 2; here we just sketch the main idea.

Let G be a group and $g \in G$. Denote by $C_G(g)$ the *centralizer* of g in G, i.e., the set of elements $h \in G$ such that $hg = gh$. For $S = \{g_1, \ldots, g_k\} \subseteq G$, $C_G(g_1, \ldots, g_k)$ denotes the centralizer of S in G, which is the intersection of the centralizers $C_G(g_i), i = 1, \ldots, k$.

Now, given a public $w \in G$, Alice privately selects $a_1 \in G$ and publishes a subgroup $B \subseteq C_G(a_1)$ (we explain why computing B is easy). Similarly, Bob privately selects $b_2 \in G$ and publishes a subgroup $A \subseteq C_G(b_2)$. Alice then selects $a_2 \in A$ and sends $w_1 = a_1 w a_2$ to Bob, while Bob selects $b_1 \in B$ and sends $w_2 = b_1 w b_2$ to Alice.

Thus, in the first transmission, say, the adversary faces the problem of finding a_1, a_2 such that $w_1 = a_1 w a_2$, where $a_2 \in A$, but there is no explicit indication of where to choose a_1 from. Therefore, before arranging something like a length attack in this case, the adversary would have to compute the centralizer $C_G(B)$ first (because $a_1 \in C_G(B)$), which is usually a hard problem by itself.

2. The protocol

In this section we give a formal description of our protocol, but first we introduce one more piece of notation. As it is common in public key exchange based on abstract groups, when transmitting an element $g \in G$ of a group, one actually uses its *normal form* $N(g)$ which is a sequence of symbols uniquely defined for a given g. A specific way of constructing such a sequence depends, of course, on a particular platform group G which we discuss in subsequent sections of our paper.

Our protocol is the following sequence of steps.

Protocol:

(1) Alice chooses an element $a_1 \in G$ of length l, chooses a subgroup of $C_G(a_1)$, and publishes its generators $A = \{\alpha_1, \ldots, \alpha_k\}$ (see the following subsection 2.1 for specifications).

(2) Bob chooses an element $b_2 \in G$ of length l, chooses a subgroup of $C_G(b_2)$, and publishes its generators $B = \{\beta_1, \ldots, \beta_m\}$ (see the following subsection 2.1 for specifications).

(3) Alice chooses a random element a_2 from $\langle \beta_1, \ldots, \beta_m \rangle$ and sends the normal form $P_A = N(a_1 w a_2)$ to Bob.

(4) Bob chooses a random element b_1 from $\langle \alpha_1, \ldots, \alpha_k \rangle$ and sends the normal form $P_B = N(b_1 w b_2)$ to Alice.

(5) Alice computes $K_A = a_1 P_B a_2$.

(6) Bob computes $K_B = b_1 P_A b_2$.

Since $a_1 b_1 = b_1 a_1$ and $a_2 b_2 = b_2 a_2$, we have $K = K_A = K_B$, the shared secret key.

2.1. Suggested values of parameters. We suggest to use the following values of parameters in the above protocol: $G = B_n$, the group of braids on n strands (see our Section 4); $n = 64$; $l = 1024$. At Step (1) of the protocol Alice generates (a_1, A) and at Step (2) Bob generates (b_2, B), both using the algorithm from [7] for computing centralizers (actually, there is no need to compute the whole centralizer, just a couple of elements are sufficient).

3. Requirements on the platform group G

In this section we discuss possible attacks on the protocol described in the previous section, and also put together some requirements on the platform group G.

To break the protocol it is sufficient to find either Alice's or Bob's private key which may be accomplished as follows:

Attack on Alice's private key. Find an element a_1' which commutes with every element of the subgroup $\langle A \rangle$ and an element $a_2' \in \langle B \rangle$, such that $P_A = N(a_1' w a_2')$. The pair (a_1', a_2') is equivalent to (a_1, a_2). (That means, $a_1' w a_2' = a_1 w a_2$, and therefore the pair (a_1', a_2') can be used by the adversary to get the shared secret key.)

Attack on Bob's private key. Find an element $b_1' \in \langle A \rangle$ and an element b_2' which commutes with every element of the subgroup $\langle B \rangle$, such that $P_B = N(b_1' w b_2')$. The pair (b_1', b_2') is equivalent to (b_1, b_2).

Consider the attack on Alice's private key (the other one is similar). The most obvious way to carry out such an attack is the following:

(A1) Compute the centralizer $C_G(A)$.
(A2) Solve the search version of the membership problem in the double coset $C_G(A) \cdot w \cdot \langle B \rangle$

To make the protocol secure, we want both these problems to be computationally hard. For the problem (A2) to be hard, it is necessary for the centralizer $C_G(A)$ to be large. Otherwise, the adversary can use the "brute force" attack, i.e., enumerate all elements of $C_G(A)$ and find candidates for b_2' (assuming that the decisional membership problem in the subgroup B is efficiently solvable).

Thus the platform group G should satisfy at least the following properties in order for our key establishment protocol to be efficient and secure.

(P1) G should be a non-commutative group of exponential growth. The latter means that the number of elements of length n in G is exponential in n; this is needed to prevent attacks by complete exhaustion of the key space.
(P2) There should be an efficiently computable normal form for elements of G.
(P3) It should be computationally easy to perform group operations (multiplication and inversion) on normal forms.
(P4) It should be computationally easy to generate pairs $(a, \{a_1, \ldots, a_k\})$ such that $aa_i = a_i a$ for each $i = 1, \ldots, k$. (Clearly, in this case the subgroup generated by $a_1, \ldots, a_k$ centralizes a).
(P5) For a generic set $\{g_1, \ldots, g_k\}$ of elements of G it should be difficult to compute
$$C(g_1, \ldots, g_n) = C(g_1) \cap \ldots \cap C(g_k).$$

(P6) Even if $H = C(g_1, \ldots, g_n)$ is computed, it should be hard to find $x \in H$ and $y \in H_1$ (where H_1 is some fixed subgroup given by a generating set) such that $xwy = w'$, i.e., to solve the membership search problem for a double coset.

4. Braid groups

In this section we consider a particular class of groups, namely braid groups, which were a popular choice for the platform of various cryptographic protocols in the last 6-7 years, starting with the seminal paper [1].

Let B_n be the group of braids on n strands and $X_n = \{x_1, \ldots, x_{n-1}\}$ the set of standard generators. Thus,

$$B_n = \langle x_1, \ldots, x_{n-1};\ x_i x_{i+1} x_i = x_{i+1} x_i x_{i+1},\ x_i x_j = x_j x_i \text{ for } |i-j| > 1 \rangle.$$

For more information on braid groups, we refer to the monographs [2], [6]; here we address the properties (P1)-(P6) from the previous section.

(P1) Braid groups B_n are non-commutative groups of exponential growth if $n \geq 3$.

(P2) There are several known normal forms for elements of B_n, including Garside normal form (see [2]) and Birman-Ko-Lee normal form [3]. Both of these forms are efficiently computable (in quadratic time with respect to the length of a given element).

(P3) There are quadratic time algorithms to multiply or invert normal forms of elements of B_n.

(P4) It is not so easy to compute the whole centralizer of an element g of G (cf. [11]). The number of steps required to compute $C_G(g)$ is proportional to $|SSS(g)|$, the size of the "super summit set" of g, which is typically huge. Nevertheless, there are approaches to finding "large parts" of $C_G(g)$, e.g. one can generate a sufficiently large part of $SSS(g)$ and pick several elements from there, see [11] for more details.

(P5) For a generic subgroup A it is hard to compute $C_G(A)$. The complexity of such computation is proportional to $|SS(A)|$, the size of the summit set of A (see [7]), which is typically huge.

(P6) There is no known solution to the membership search problem for double cosets $H \cdot w \cdot H'$ in braid groups. This problem, in theory, appears to be much more complicated (for generic subgroups H and H') than the conjugacy search problem.

5. Generating commuting elements in braid groups

In this section, we explain how one can efficiently generate elements commuting with a given element of a braid group, which is important for our protocol (steps (1), (2)).

In a group B_n, for $i < j$ define a braid word $\delta_{i,j} = x_i \ldots x_j$. Denote $\delta_{1,n-1}$ by just δ. Clearly for any braid word $w = x_{i_1}^{\varepsilon_1} \ldots x_{i_k}^{\varepsilon_k}$ such that $i_j < n - 1 - m$ for each $j = 1, \ldots, k$, one has

$$w^{\delta^m} = x_{i_1+m}^{\varepsilon_1} \ldots x_{i_k+m}^{\varepsilon_k} \text{ in } B_n.$$

Now given $w \in B_n$, we are going to generate a set C of elements commuting with w, as follows.

A. **Initial setup:**
- Take $n = 32$.
- Generate w_1 to be a random freely reduced braid word of length 40 in the generators $\{x_1, x_2, x_3\}$.
- For $j = 2, \ldots, 8$ generate random freely reduced braid words c_i of length 20 in the generators $\{x_1, x_2, x_3\}$ and put $w_j = c_j^{-1} w_1 c_j$. Let c_1 be ε, the empty word.
- Let $w = w_1 w_2^{\delta^4} w_3^{\delta^8} \ldots w_8^{\delta^{28}}$.

- Compute generators of the centralizer $C(w_1)$ (using technique from [**11**]). Clearly, $C(w_i) = C(w_1)^{c_i}$ for $i = 2, \ldots, 8$. Let
$$C = C(w_1) \cup C(w_1)^{\delta^4 c_2} \cup \ldots \cup C(w_1)^{\delta^{28} c_8}.$$

- Add to C braid words $flip(1,2), \ldots, flip(7,8)$, where
$$flip(i, i+1) = (\delta_{4i,4i-3}\delta_{4i+1,4i-2}\delta_{4i+2,4i-1}\delta_{4i+3,4i})^2.$$
These commute with w.

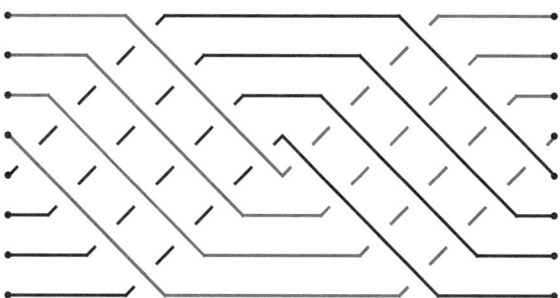

FIGURE 1. Flip.

- Add to C braid words $swap(1,2), \ldots, swap(7,8)$, where $swap(i, i+1)$ is the following braid:
$$(c_i^{-1} c_{i+1})^{\delta^{(4i-4)}} (c_{i+1}^{-1} c_i)^{\delta^{(4i)}} \delta_{4i,4i-3}\delta_{4i+1,4i-2}\delta_{4i+2,4i-1}\delta_{4i+3,4i}.$$
These, too, commute with w.

B. **For more diffusion:** Generate a random braid word x of length 150 in the generators of B_n and conjugate w as well as elements of C by x.

The word w generated this way has length bounded by $8 \cdot 40 + 2 \cdot 150 = 620$. Also note that $|flip(i, i+1)| = 32$ and $|swap(i, i+1)| \leq 20 + 20 + 20 + 20 + 16 = 96$. We can therefore select sufficiently many elements of length ≤ 100 from the set C and publish them (in Garside normal form).

6. Semantic security

In this section, we discuss *semantic security* of a cryptosystem that would be based on a shared key obtained in our protocol. Semantic security is the standard notion of security for encryption protocols, see [**10**].

Security of the protocol described in our Section 2 is based on the assumption that the following problem is computationally hard:

Given the public information w, P_A, and P_B it is hard to compute the shared key K.

This assumption is the *computational assumption of the protocol*. The stronger *decisional* version of this assumption would be:

Given w, P_A, and P_B, it is hard to distinguish the shared key K from a random element of the form awb.

We should point out that without this decisional assumption, it may still be possible to design a semantically secure encryption protocol in the "random oracle model" the same way it was done in [**15**, Section 3.3], namely, by employing a hash function $H : B_n \to \{0,1\}^k$ from the braid group to the message space. Still, it would be quite interesting to find out whether or not the shared key K obtained in our key establishment protocol can be directly used for semantically secure encryption.

The decisional assumption above appears to be wrong for most choices of w, P_A and P_B because of the following consideration. Since $P_A = a_1 w a_2$, we have $a_1 = P_A a_2^{-1} w^{-1}$. Therefore, $K = a_1 b_1 w b_2 a_2 = P_A a_2^{-1}(w^{-1} P_B) a_2$. Hence, K is a product of a public element P_A and a public element $w^{-1} P_B$ conjugated by an element from a subgroup $\{\beta_1, \ldots, \beta_k\}$.

It seems plausible that, for some choices of the keys, elements of this type can be distinguished from random elements of the form awb along the same lines it was done in [9] (in a different, but similar context). Indeed, if $w^{-1} P_B$ is not a *pure braid*, then it projects to a non-trivial permutation, call it ρ_B, under the natural homomorphism π from the braid group B_n onto the symmetric group S_n. Then the conjugate permutation $\pi(a_2)^{-1} \rho_B \pi(a_2)$ has the same cyclic structure as ρ_B does, and this gives away some information about the permutation $\pi(K) = \pi(P_A) \pi(a_2)^{-1} \rho_B \pi(a_2)$; for example, from knowing $\pi(P_A)$ and the cyclic structure of $\pi(a_2)^{-1} \rho_B \pi(a_2)$, one can get information about possible order of the permutation $\pi(K)$.

If both P_A and $w^{-1} P_B$ are pure braids, then it is possible to use other homomorphisms (e.g. pulling out a strand) to obtain some partial information; see [9] for details. If $w^{-1} P_B$ is a pure braid but P_A is not, then, again, the homomorphism π reveals partial information about the shared key K.

Acknowledgement. The second author is grateful to Alexei G. Myasnikov and Robert Gilman for their kind support and encouragement.

References

[1] I. Anshel, M. Anshel, D. Goldfeld, *An algebraic method for public-key cryptography*, Math. Res. Lett. **6** (1999), 287–291.

[2] J. S. Birman, *Braids, links and mapping class groups*, Ann. Math. Studies **82**, Princeton Univ. Press, 1974.

[3] J. S. Birman, K. H. Ko, S. J. Lee, *A new approach to the word and conjugacy problems in the braid groups*, Adv. Math. **139** (1998), 322–353.

[4] F. Celler, C. Leedham-Green, S. H. Murray, A. Niemeyer, E. A. O'Brien, *Generating random elements of a finite group*, Comm. Algebra **23** (1995), 4931–4948.

[5] J. C. Cha, K. H. Ko, S. J. Lee, J. W. Han, J. H. Cheon, *An Efficient Implementation of Braid Groups*, ASIACRYPT 2001, Lecture Notes in Comput. Sci. **2248** (2001), 144–156.

[6] D. B. A. Epstein, J. W. Cannon, D. F. Holt, S. V. F. Levy, M. S. Paterson, W. P. Thurston, *Word processing in groups*. Jones and Bartlett Publishers, Boston, MA, 1992.

[7] N. Franco, J. Gonzalez-Meneses, *Computation of Centralizers in Braid groups and Garside groups*, Rev. Mat. Iberoamericana **19** (2) (2003), 367–384.

[8] D. Garber, S. Kaplan, M. Teicher, B. Tsaban, U. Vishne, *Probabilistic solutions of equations in the braid group*, Advances in Applied Mathematics **35** (2005), 323–334.

[9] R. Gennaro and D. Micciancio, *Cryptanalysis of a pseudorandom generator based on braid groups*, in EUROCRYPT 2002, Lecture Notes in Comput. Sci. **2332** (2002), 1–13.

[10] S. Goldwasser and S. Micali, *Probabilistic encryption*, Journal of Computer and System Sciences **28** (1984), 270–299.

[11] J. Gonzalez-Meneses and B. Wiest, *On the structure of the centraliser of a braid*, Ann. Sci. École Norm. Sup. **37** (5) (2004), 729–757.

[12] D. Hofheinz and R. Steinwandt, *A practical attack on some braid group based cryptographic primitives*, in Public Key Cryptography, 6th International Workshop on Practice and Theory in Public Key Cryptography, PKC 2003 Proceedings, Y.G. Desmedt, ed., Lecture Notes in Computer Science **2567**, pp. 187–198, Springer, 2002.

[13] J. Hughes and A. Tannenbaum, *Length-based attacks for certain group based encryption rewriting systems*, Workshop SECI02 Securitè de la Communication sur Intenet, September 2002, Tunis, Tunisia.
http://www.storagetek.com/hughes/

[14] D. Grigoriev, I. Ponomarenko, *Homomorphic public-key cryptosystems and encrypting boolean circuits*, preprint.
http://eprint.iacr.org/2003/025
[15] K. H. Ko, S. J. Lee, J. H. Cheon, J. W. Han, J. Kang, C. Park, *New public-key cryptosystem using braid groups*, Advances in cryptology—CRYPTO 2000 (Santa Barbara, CA), 166–183, Lecture Notes in Comput. Sci. **1880**, Springer, Berlin, 2000.
[16] A. G. Myasnikov, V. Shpilrain, and A. Ushakov, *A practical attack on some braid group based cryptographic protocols*, in CRYPTO 2005, Lecture Notes Comp. Sci. **3621** (2005), 86–96.
[17] V. Shpilrain and A. Ushakov, *Thompson's group and public key cryptography*, Lecture Notes Comp. Sc. **3531** (2005), 151–164.
[18] V. Shpilrain and G. Zapata, *Combinatorial group theory and public key cryptography*, Applicable Algebra in Engineering, Communication and Computing, to appear.
http://eprint.iacr.org/2004/242
[19] V. Shpilrain and G. Zapata, *Using the subgroup membership search problem in public key cryptography*, this volume.

DEPARTMENT OF MATHEMATICS, THE CITY COLLEGE OF NEW YORK, NEW YORK, NY 10031
E-mail address: shpil@groups.sci.ccny.cuny.edu

DEPARTMENT OF MATHEMATICS, CUNY GRADUATE CENTER, NEW YORK, NY 10016
E-mail address: aushakov@mail.ru

Using the subgroup membership search problem in public key cryptography

Vladimir Shpilrain and Gabriel Zapata

ABSTRACT. There are several public key protocols around that use the computational hardness of either the *conjugacy search problem* or the *word (search) problem* for nonabelian groups. In this paper, we describe a cryptosystem whose security is based on the computational hardness of the *subgroup membership (search) problem*: given a group G, a subgroup H generated by $h_1, \ldots, h_k$, and an element $h \in H$, find an expression of h in terms of $h_1, \ldots, h_k$.

It is also interesting to note that groups which we suggest to use as the platform, *free metabelian groups*, are *infinitely presented*, in contrast with groups typically used in public key cryptography. Nevertheless, these group have efficiently (and, in fact, very easily) solvable word problem.

1. Introduction

In search for a more efficient and/or secure alternative to established cryptographic protocols (such as RSA), several authors have come up with public key establishment protocols as well as with complete public key cryptosystems based on allegedly hard *search problems* from combinatorial (semi)group theory, including the conjugacy search problem [1, 16], the homomorphism search problem [11], [12], [24], the decomposition search problem [5, 16, 22].

In this paper, we describe a cryptosystem whose security is based on the computational hardness of the *subgroup membership (search) problem*:

Given a group G, a subgroup H generated by $h_1, \ldots, h_k$, and an element $h \in H$, find an expression of h in terms of $h_1, \ldots, h_k$.

It should be mentioned that, in fact, the Anshel-Anshel-Goldfeld protocol which may seem to rely on the computational hardness of the conjugacy search problem alone, actually relies (perhaps somewhat implicitly) on the hardness of the subgroup membership (search) problem as well, as explained in [22]. We also note that there is some similarity, at a philosophical level, between our cryptosystem and homomorphic public-key cryptosystems of [11] and [12]. However, there are also essential differences, as explained in our Section 2.

First we outline the ideas behind our cryptosystem. These ideas can be traced back to [19], where a similar approach was used in commutative situation, but the resulting cryptosystem was not accepted by the cryptographic community as secure [10].

2000 *Mathematics Subject Classification.* 94A60, 20E05, 20F05, 68P25.
Research of the first author was partially supported by the NSF grant DMS-0405105.

We believe that employing a nonabelian group instead of a polynomial algebra as the platform does make a difference in both the efficiency and security components. We note that various nonabelian groups have been used as platforms in the above mentioned protocols; in particular, braid groups [1], [16], Grigorchuk groups [9], Thompson's group [22], polycyclic groups [7]. In this paper, we use groups of a different nature.

To explain our approach, we need some more background. Let F_n denote the free group of rank n. It is well known that any map on the generators of F_n into F_n extends to an endomorphism of F_n. It is also well known that free groups are not the only groups with this property.

DEFINITION 1. Let F be a free group and let R be a normal subgroup of F. The factor group F/R is called *relatively free* if R is fully invariant, i.e., if $\alpha(R) \leq R$ for any endomorphism α of F. If $x_1, \ldots, x_n$ are free generators of F, then $x_1 R, \ldots, x_n R$ are called relatively free generators of F/R.

We are going to denote relatively free generators of F/R simply by $x_1, \ldots, x_n$ when there is no ambiguity. Let $\mathcal{F}_n$ denote a relatively free group of rank n, i.e., $\mathcal{F}_n = F_n/R$ for some fully invariant R. Then any map on its generators into $\mathcal{F}_n$ can be extended to an endomorphism of $\mathcal{F}_n$. Because of this property, any relatively free group $\mathcal{F}_n$ can be a candidate for the platform of our cryptosystem.

We also need to recall one basic fact about automorphisms of F_n. Let $X = \{x_1, \ldots, x_n\}$ be a set of generators of F_n and consider the maps $\alpha_j, \beta_{jk} : X \to F_n$ given by

$$\alpha_j : x_i \longmapsto \begin{cases} x_i^{-1} & \text{if } i = j \\ x_i & \text{if } i \neq j \end{cases} \quad \text{and} \quad \beta_{jk} : x_i \longmapsto \begin{cases} x_i x_j & \text{if } i = k \\ x_i & \text{if } i \neq k, \end{cases}$$

where $1 \leq i, j, k \leq n$. Then all the α_j's and β_{jk}'s define automorphisms of F_n, which are called *Nielsen automorphisms*, and generate the whole automorphism group of F_n.

Nielsen automorphisms can be defined for any relatively free group $\mathcal{F}_n$ the same way. They generate a subgroup of the group $Aut(\mathcal{F}_n)$ of all automorphisms of $\mathcal{F}_n$; this subgroup is called the group of *tame* automorphisms. In some cases, it is equal to the whole $Aut(\mathcal{F}_n)$ (see e.g. [2]); in other cases, it is a proper subgroup of $Aut(\mathcal{F}_n)$ (see e.g. [4, 21]). For our purposes, it is important that groups of the form $F_n/[R, R]$ (where R is a normal subgroup of F_n, and $[R, R]$ its commutator subgroup) tend to have many non-tame automorphisms by a result of [21]; we reproduce it in our Section 3.

To conclude the introduction, we introduce some more notation. Let $\langle x_1, \ldots, x_n, x_{n+1}, \ldots, x_{n+m} ; R \rangle$ be a presentation of a relatively free group $\mathcal{F}_{n+m}$. Let $\varphi \in Aut \, \mathcal{F}_{n+m}$ be an automorphism such that

$$\varphi : x_i \longmapsto y_i, \quad \text{where} \quad y_i = y_i(x_1, \ldots, x_{n+m}).$$

This φ acts on $\mathcal{F}_{n+m}^{n+m}$, the direct product of $n + m$ copies of $\mathcal{F}_{n+m}$, in the natural way:

$$\varphi : (x_1, \ldots, x_{n+m}) \longmapsto (y_1, \ldots, y_{n+m}),$$

and, more generally,

$$\varphi : (u_1, \ldots, u_{n+m}) \longmapsto (y_1(u_1, \ldots, u_{n+m}), \ldots, y_{n+m}(u_1, \ldots, u_{n+m}))$$

for any $u_i = u_i(x_1, \ldots, x_{n+m}) \in \mathcal{F}_{n+m}$.

Then, let $\hat{\varphi}$ be the restriction of φ to the subgroup of $\mathcal{F}_{n+m}^{n+m}$ that consists of all elements of the form $(u_1(x_1, \ldots, x_n), \ldots, u_n(x_1, \ldots, x_n), 1, \ldots, 1)$ (this subgroup is isomorphic to $\mathcal{F}_n^n$), so that

$$\hat{\varphi} : (u_1, \ldots, u_n, 1, \ldots, 1) \longmapsto (\hat{y}_1(u_1, \ldots, u_n), \ldots, \hat{y}_{n+m}(u_1, \ldots, u_n)),$$

where $\hat{y}_i(x_1, \ldots, x_n) = y_i(x_1, \ldots, x_n, 1, \ldots, 1)$.

We note that $\hat{\varphi}$ is a one-to-one map because the restriction of a one-to-one map to a subgroup is itself one-to-one. This property will be important to us in Section 2. At the same time, there is no visible way of recovering φ from $\hat{\varphi}$.

2. The protocol

Let $\mathcal{F}_{n+m}$ be a relatively free group. We discuss a specific choice of the platform group in the following Section 3; here we present our public key encryption/decryption protocol without specifying the platform group.

Key Generation: (1) Alice privately chooses an automorphism $\varphi \in Aut\, \mathcal{F}_{n+m}$ as a random product of Nielsen automorphisms and some easily invertible automorphisms of the type given in the theorem in the Introduction: $\varphi = \tau_1 \cdots \tau_k$ and computes the inverse $\varphi^{-1} = \tau_k^{-1} \cdots \tau_1^{-1}$. This φ^{-1} is Alice's private decryption key.

(2) Let $y_i = y_i(x_1, \ldots, x_{n+m}) = \varphi(x_i)$. Alice publishes $\hat{y}_i(x_1, \ldots, x_n) = y_i(x_1, \ldots, x_n, 1, \ldots, 1)$, $i = 1, \ldots, n+m$. This collection of $\hat{y}_i$ is the public encryption key.

Encryption and Decryption: The information considered for encryption (i.e., Bob's private message) is an element w of the subgroup of $\mathcal{F}_{n+m}^{n+m}$ that consists of all elements of the form $(u_1(x_1, \ldots, x_n), \ldots, u_n(x_1, \ldots, x_n), 1, \ldots, 1)$. Thus,

$$w = (w_1(x_1, \ldots, x_n), \ldots, w_n(x_1, \ldots, x_n))$$

(one can suppress the 1's in the end).

(1) Encryption is defined by

$$w \mapsto \hat{\varphi}(w) = (\hat{y}_1(w_1, \ldots, w_n), \ldots, \hat{y}_{n+m}(w_1, \ldots, w_n)).$$

Bob then transmits $\hat{\varphi}(w)$ as a tuple of words in the alphabet X.

(2) Decryption is defined by $\hat{\varphi}(w) \mapsto [\varphi^{-1}(\hat{\varphi}(w))]\,|_{x_{n+1}=\ldots=x_{n+m}=1} = w$. This w (more precisely, a *normal form* of w, see the end of Section 3) is Alice's and Bob's common secret key.

We note that the encryption of w here is *not* a homomorphic image (or a preimage) of w, in contrast to homomorphic public-key cryptosystems of [11] and [12].

The adversary can recover the plaintext w if he finds an expression for each $x_1, \ldots, x_n$ in terms of $\hat{y}_1, \ldots, \hat{y}_{n+m}$, i.e., if he solves the membership search problem for the subgroup generated by $\hat{y}_1, \ldots, \hat{y}_{n+m}$ (this subgroup is actually $\mathcal{F}_n$). Then from

$$x_i = u_i(\hat{y}_1(x_1, \ldots, x_n), \ldots, \hat{y}_n(x_1, \ldots, x_n))$$

he can get

$$w_i = u_i(\hat{y}_1(w_1, \ldots, w_n), \ldots, \hat{y}_n(w_1, \ldots, w_n)).$$

If however the adversary wants to completely break the cryptosystem, i.e., to get the private decryption key, then he has to find the automorphism φ (and its inverse) based on the restriction $\hat{\varphi}$. This problem looks unapproachable to us by any deterministic method other than trying out products of "elementary" automorphisms of $\mathcal{F}_{n+m}$ until the one with the right restriction is found. This method is computationally infeasible for large n, m.

We discuss specific values of the parameters of our cryptosystem in Section 5. Here we give a "toy example" just to illustrate how our protocol works. This example is not platform-specific because we only use Nielsen automorphisms as building blocks here.

EXAMPLE 1. Let $n = 2, m = 1$, and let β_{jk} be Nielsen automorphisms of $\mathcal{F}_3$ defined in the Introduction. Let $\varphi = \beta_{23}^2 \beta_{12} \beta_{31} \beta_{32}$. Then $\varphi : x_1 \to x_1 x_3 x_2^2$, $x_2 \to x_2 x_1 x_3 x_2^2$, $x_3 \to x_3 x_2^2$. Thus, $y_1 = x_1 x_3 x_2^2$, $y_2 = x_2 x_1 x_3 x_2^2$, $y_3 = x_3 x_2^2$, and therefore $\hat{y}_1 = x_1 x_2^2$, $\hat{y}_2 = x_2 x_1 x_2^2$, $\hat{y}_3 = x_2^2$.

Thus, Bob's private message $(u_1(x_1, x_2), u_2(x_1, x_2))$ is encrypted as $(u_1 u_2^2, u_2 u_1 u_2^2, u_2^2) = (v_1, v_2, v_3)$.

In this simple example the adversary can easily obtain x_1, x_2 from the public $\hat{y}_i$ as follows: $x_1 = \hat{y}_1 \hat{y}_3^{-1}$, $x_2 = \hat{y}_2 \hat{y}_1^{-1}$. Therefore, he can recover $u_1 = v_1 v_3^{-1}$, $u_2 = v_2 v_1^{-1}$.

As a comment to this example, we note that in a free group, the subgroup membership search problem can be efficiently solved by Nielsen's method (see e.g. [**17**]). However, adding automorphisms $\alpha_{u,j}$ described in the Theorem in the next Section 3 makes a difference because it makes working in a free group pointless. We illustrate this by Example 2 in the next section.

3. Free metabelian groups

In this section, we suggest specific relatively free groups, called *free metabelian groups*, which can be used as platforms for the cryptosystem described in the previous section. Some terminology has to be introduced first.

A group G is called *abelian* (or commutative) if $[a, b] = 1$ for any $a, b \in G$, where $[a, b]$ is the notation for $a^{-1}b^{-1}ab$. This can be generalized in different ways. A group G is called *metabelian* if $[[x, y], [z, t]] = 1$ for any $x, y, z, t \in G$. A group G is called *nilpotent of class* $c \geq 1$ if $[y_1, y_2, \ldots, y_{c+1}] = 1$ for any $y_1, y_2, \ldots, y_{c+1} \in G$, where $[y_1, y_2, y_3] = [[y_1, y_2], y_3]$, etc.

The commutator subgroup of G is the group $G' = [G, G]$ generated by all commutators, i.e., by expressions of the form $[u, v] = u^{-1}v^{-1}uv$, where $u, v \in G$. Furthermore, we can define, by induction, the kth term of the *lower central series* of G: $\gamma_1(G) = G, \gamma_2(G) = [G, G], \gamma_k(G) = [\gamma_{k-1}G, G]$. Note that one has $\alpha([u, v]) = [\alpha(u), \alpha(v)]$ for any endomorphism α of G. Therefore, $\gamma_k(G)$ is a fully invariant subgroup of G for any $k \geq 1$, and so is $G'' = [G', G']$.

DEFINITION 2. Let F_n be the free group of rank n. The relatively free group F_n/F_n'' is called the *free metabelian group* of rank n, which we denote by M_n.

Our choice of free metabelian groups for the platform is motivated by the following facts:

(1) The word problem in the group M_n is solvable in time $O(m^2 n)$ with respect to the length m of a given word. Moreover, every element of M_n has a unique associated "normal form", which makes it possible to use our protocol for an efficient encryption/decryption of actual messages without prior common key establishment.

(2) M_n has exponential growth which provides for an exponential (with respect to the key size) key space.

(3) Subgroup membership (search) problem in M_n has no known polynomial-time solution.

Here (2) is a well-established fact [**18**], so we leave it without further comments. As for (3), there are two different solutions of the membership (decision) problem in M_n known by now [**6**, **20**], but none of them is "even close" to yielding a polynomial-time algorithm for solving the membership search problem. We give more details in Section 5; here we concentrate on the word problem in M_n.

It is also interesting to note that free metabelian groups M_n are infinitely presented if $n \geq 2$ (see e.g. [**3**]), i.e., they cannot be defined by finitely many relations, in contrast with groups typically used in public key cryptography. Nevertheless, as we shall see in the next section, these groups have efficiently (and, in fact, very easily) solvable word problem.

Now we reproduce a result from [21] which provides us with "non-standard" automorphisms of free metabelian groups.

Theorem. [21] Let $\mathcal{F}_n = F_n/[R,R]$, and let $u \in R$. If $R \leq \gamma_3(F_n)$ and $n \geq 2$, then the following automorphism $\alpha_{u,j}$ of $\mathcal{F}_n$ is not tame: $\alpha_{u,j} : x_j \to x_j[x_j, u, x_j]$, $x_i \to x_i$, $i \neq j$.

Note that $\alpha_{u,j}^{-1} : x_j \to x_j[x_j, u^{-1}, x_j]$, $x_i \to x_i$, $i \neq j$. Also note that this theorem cannot be applied to free metabelian groups $M_n = F_n/[F_n', F_n']$ because $F_n' \not\leq \gamma_3(F_n)$. In fact, it is known that free metabelian groups M_n have only tame automorphisms if $n \neq 3$ [2]. However, automorphisms $\alpha_{u,j}$ described in the Theorem above, although tame in the case of a free metabelian group, are tame in a very nontrivial ("nonmonotonic") way, i.e., to factor them into a product of Nielsen automorphisms one has to first increase the length of $\alpha_{u,j}(x_i)$ before it could be decreased. This is supposed to foil "length based" attacks on φ (see Section 5 for more details).

To conclude this section, we modify Example 1 from the previous section by composing the automorphism φ from that example with the map $\alpha = \alpha_{[x_1,x_2],1}$ which is an automorphism of a free metabelian group, but not of a free group.

EXAMPLE 2. Let $n = 2, m = 1$, and let β_{jk} be Nielsen automorphisms of $\mathcal{F}_3$ defined in the Introduction. Let $\varphi = \beta_{23}^2 \beta_{12} \beta_{31} \beta_{32}$. Then $\varphi : x_1 \to x_1 x_3 x_2^2$, $x_2 \to x_2 x_1 x_3 x_2^2$, $x_3 \to x_3 x_2^2$. Now compose φ with the map $\alpha : x_1 \to x_1[x_1, [x_1, x_2], x_1]$, $x_i \to x_i$, $i \neq 1$. The resulting automorphism takes x_1 to $y_1 = x_1[x_1, [x_1, x_2], x_1]x_3 x_2^2$, x_2 to $y_2 = x_2 x_1[x_1, [x_1, x_2], x_1]x_3 x_2^2$, and x_3 to $y_3 = x_3 x_2^2$, and therefore $\hat{y}_1 = x_1[x_1, [x_1, x_2], x_1]x_2^2$, $\hat{y}_2 = x_2 x_1[x_1, [x_1, x_2], x_1]x_2^2$, $\hat{y}_3 = x_2^2$.

Now in the free group F_2, the elements x_1 and x_2 no longer belong to the subgroup generated by $\hat{y}_1, \hat{y}_2, \hat{y}_3$. At the same time, since α is an automorphism of the free metabelian group F_3/F_3'', x_1 and x_2, considered as elements of F_3/F_3'', do belong to the subgroup of F_3/F_3'' generated by $\hat{y}_1, \hat{y}_2, \hat{y}_3$. The adversary will therefore have to solve the subgroup membership search problem in a free metabelian group, which is much more difficult than to do that in a free group.

4. Normal forms in free metabelian groups

In this section, we describe two different normal forms for elements of a free metabelian group M_n. The first one is good for transmissions, because it is easily convertible back to a word representing a transmitted element. However, this normal form is not unique if $n > 2$, so it cannot be used as a shared secret by Alice and Bob. For the latter purpose, the other normal form (a 2×2 matrix) can be used.

Let $u \in M_n$. By u_{ab} we denote the abelianization of u, i.e., the image of u under the natural epimorphism $\alpha : M_n \to M_n/[M_n, M_n]$. Note that we can identify $M_n/[M_n, M_n]$ with $F_n/[F_n, F_n]$. Technically, u_{ab} is an element of a factor group of F_n, but we also use the same notation u_{ab} for any word in the generators x_i (i.e., an element of the ambient free group F_n) representing u_{ab} when there is no ambiguity.

For $u, v \in M_n$, by u^v we denote the expression $v^{-1}uv$; we also say that v acts on u by conjugation. If $u \in [M_n, M_n]$, then this action can be extended to the group ring $\mathbb{Z}(M_n/[M_n, M_n])$ which we are going to denote by $\mathbb{Z}A_n$, to simplify the notation. (Here $A_n = M_n/[M_n, M_n]$ is the free abelian group of rank n.) Let $W \in \mathbb{Z}A_n$ be expressed in the form $W = \sum a_i v_i$, where $a_i \in \mathbb{Z}$, $v_i \in A_n$. Then by u^W we denote the product

$\prod (u^{a_i})^{v_i}$. This product is well-defined since any two elements of $[M_n, M_n]$ commute in M_n.

Now let $u \in M_n$. Then u can be written in the following normal form:

$$u = u_{ab} \cdot \prod_{i<j} [x_i, x_j]^{W_{ij}} \tag{1}$$

where $W_{ij} \in \mathbb{Z}A_n$. To get to this form, one can use a "collecting process" based on the following identities (recall that $[x, y] = x^{-1}y^{-1}xy$):

$$[y, x] = [x, y]^{-1}$$

$$xy = yx[x, y]$$

$$xy^{-1} = y^{-1}[y, x]^{y^{-1}x^{-1}} x$$

$$x^{-1}y = y[y, x]^{y^{-1}x^{-1}} x^{-1}$$

$$[x, y]z = z[x, y]^z.$$

The collecting process itself is simple:

(1) Using the above identities, go left to right along the word u collecting all "non-commutator" occurrences of x_1 on the left (that means, do not worry about occurrences of the form $[x_1, x_j]$ or $[x_j, x_1]$ created in the process). Repeat this with x_2, x_3, etc. In the end of this process, u will be written in the form $u = u_{ab} \cdot c$, where $c \in [M_n, M_n]$ is a product of expressions of the form $[x_i, x_j]^g$, $g \in M_n$.

(2) Since any two elements of $[M_n, M_n]$ commute in M_n, one can now easily regroup the expressions $[x_i, x_j]^g$ so that u takes the form $u = u_{ab} \cdot \prod_{i<j} [x_i, x_j]^{W_{ij}}$, where $W_{ij} \in \mathbb{Z}A_n$.

This process apparently takes quadratic time with respect to the length of u.

To convert the normal form (1) to a word is trivial because (1) is, in fact, already a word. The only problem with (1) is that it is not unique if $n > 2$, so it cannot be used as a shared secret by Alice and Bob. For the latter purpose, we are now going to introduce another normal form which is unique, efficiently computable (in quadratic time with respect to the length of u), but not so easily convertible back to a word.

We have to first introduce *Fox derivatives*, which are noncommutative analogs of usual Leibniz derivatives.

DEFINITION 3. Let $\mathbb{Z}F$ be the group ring of a free group F generated by $x_1, x_2, \ldots$. A *Fox derivation* with respect to x_i is a map $\partial_{x_i} : \mathbb{Z}F \to \mathbb{Z}F$ such that $\partial_{x_i}(x_j) = \delta_{ij}$ and $\partial_{x_i}(vw) = \partial_{x_i}(v) + v \cdot \partial_{x_i}(w)$ for any $v, w \in F$. This map can be extended to the whole $\mathbb{Z}F$ by linearity.

EXAMPLE 3. Let $g \in F$ and let 1 be the identity of F. Since $\partial(1) = \partial(1) + \partial(1)$, it follows that $\partial(1) = 0$. Therefore $\partial(gg^{-1}) = \partial(g) + g\partial(g^{-1}) = 0$, which implies $\partial(g^{-1}) = -g^{-1}\partial(g)$.

EXAMPLE 4. Let x and y be generators of the free group $F(x, y)$. Then

$$\begin{aligned}\partial_x([x, y]) = \partial_x(x^{-1}y^{-1}xy) &= \partial_x(x^{-1}) + x^{-1}\partial_x(y^{-1}) + x^{-1}y^{-1}\partial_x(x) + x^{-1}y^{-1}x\partial_x(y) \\ &= -x^{-1} + x^{-1}y^{-1} = x^{-1}y^{-1}(1 - y).\end{aligned}$$

$$\begin{aligned}\partial_y([x,y]) &= \partial_y(x^{-1}) + x^{-1}\partial_y(y^{-1}) + x^{-1}y^{-1}\partial_y(x) + x^{-1}y^{-1}x\partial_y(y) \\ &= -x^{-1}y^{-1} + x^{-1}y^{-1}x = -x^{-1}y^{-1}(1-x)\,.\end{aligned}$$

Let F_{ab} denote the abelianization of a free group F, i.e., the factor group F/F'. Let $\alpha : F \to F_{ab}$ be the natural epimorphism; it can be extended to the map $\alpha : \mathbb{Z}F \to \mathbb{Z}F_{ab}$ by linearity. A proof of the following proposition can be found in [13].

PROPOSITION 1. Let $w \in F_n$. Then $w \in F_n''$ if and only if $\alpha(\partial_{x_i}(w)) = 0$ for each generator x_i of F_n.

This proposition yields a simple algorithm for solving the word problem in a free metabelian group M_n: given $w \in M_n$ as a word in relatively free generators x_i, one considers w an element of the free group F_n with the same set of free generators, computes $\partial_{x_i}(w)$ for each x_i, and checks whether or not all of them abelianize to 0. The latter is straightforward since the word problem in the free abelian group F_{ab} is easily solvable.

This algorithm is not only simple but efficient, too:

PROPOSITION 2. The algorithm for solving the word problem in M_n based on Proposition 1 has at most quadratic time complexity with respect to the length of the input word.

PROOF. Let $w \in F_n$ and let $|w| = m$ denote the usual lexicographic length of the word w. The computation of $\partial_{x_i}(w)$, for any generator x_i, produces at most m summands in the free group ring $\mathbb{Z}F_n$. Thus, the computation of ∂_x has at most linear time complexity with respect to m. Then, deciding whether or not the abelianization of $\partial_{x_i}(w)$ is 0 amounts to collecting summands of the form $c \cdot h_i$, $c \in \mathbb{Z}, h_i \in F_n$, such that all h_i have the same abelianization. This is achieved by rewriting every h_i in the form $x_1^{a_1} x_2^{a_2} \cdot \ldots \cdot x_n^{a_n} \cdot u_i$, where $u_i \in F_n'$. Since any h_i has length $\leq m$ and the number of h_i is at most m, this part of the algorithm takes time $O(m^2)$, which completes the proof. □

Finally, we describe the normal form of $u \in M_n$ based on Fox derivatives.

For an element $u \in M_n$ of a free metabelian group, its normal form is a 2×2 matrix with the following entries:

(1) The entry in the lower left corner is 0
(2) The entry in the lower right corner is 1
(3) The entry in the upper left corner is the abelianization of u, so it is an element of the free abelian group $M_n/[M_n, M_n]$
(4) The entry in the upper right corner is the most essential one. It is a vector of n abelianized partial Fox derivatives of the word u.

We note that the free abelian group $M_n/[M_n, M_n]$ acts on vectors of abelianized Fox derivatives by (componentwise) multiplication. This makes the set of normal forms a group under multiplication. Furthermore, the representation of elements of M_n by their normal forms is faithful, i.e., we actually have an embedding of M_n into a group of matrices; this is called *Magnus embedding* (see e.g. [13]).

5. Parameters and cryptanalysis

There are two visible ways to attack our cryptosystem: (1) trying to find the automorphism φ (and its inverse) based on the restriction $\hat{\varphi}$ and (2) trying to solve the subgroup membership (search) problem in the platform group. As we have already pointed out in Section 2, the first way looks intractable to us, so we focus on the subgroup membership problem here, but first we make a relevant remark about the word problem.

In the previous section, we obtained the solvability of the word problem in free metabelian groups M_n by using Fox derivations. Here we note that the solvability of the word problem in M_n can also be obtained by a much less efficient but nevertheless interesting method, using the fact that M_n are *residually finite groups* [14].

DEFINITION 4. A group G is said to be *residually finite* if for every $g \in G - \{1\}$ there exists a finite group H_g and a homomorphism $\varphi : G \to H_g$ such that $\varphi(g) \neq 1$.

If a group $G = F/R$ is residually finite, then it has the word problem solvable as follows. Given $g \in G$ as a word in the generators of F, two algorithms run in parallel. One of them goes over all finite homomorphic images of G (those are recursively enumerable) and checks whether $\varphi(g) \neq 1$ in any of them (note that the word problem is solvable in any finite group). Thus, if $g \neq 1$ in G, this algorithm will eventually detect that. The other algorithm just recursively enumerates all elements of R and compares them to g one at a time. Thus, if $g = 1$ in G, this algorithm will eventually detect that.

This way of solving the word problem is "cute", but it is obviously useless in real life. We have mentioned it here only because there is a similar (equally useless) way of solving the membership problem in a special class of groups.

DEFINITION 5. A group G is said to be *locally extended residually finite* (LERF) if for every subgroup $K \in G$ and $g \in G - K$, there exists a finite group H and a homomorphism $\varphi : G \to H$ such that $\varphi(g) \in \varphi(G) - \varphi(K)$.

Sometimes, LERF groups are also called *subgroup separable*.

Coulbois [6] showed that free metabelian groups are LERF. If a group G is LERF, it follows that G has solvable subgroup membership (decision) problem by way of an algorithm similar to what was described above. However, as well as the algorithm for solving the word problem based on residual finiteness of a given group, this algorithm is anything but practical.

An alternative algorithm for solving the subgroup membership (both decision and search) problem in M_n was offered in [20]. This algorithm is somewhat more "down-to-earth", but it is still far from being practical because it involves, among other things, computing a Gröbner basis of an ideal of the ring of Laurent polynomials in n variables.

In the absence of feasible deterministic attacks, one might try heuristic ("length based") attacks in the spirit of [8, 15]. We point out however that groups that were typically used as proving ground for "length based" attacks are very different in nature from free metabelian groups, and the length function for free metabelian groups is, informally speaking, much less predictable than it is, say, for braid groups, which probably means that "length based" attacks are going to be less successful.

We are now going to discuss specific parameters of our cryptosystem. First, we suggest to use a free metabelian group as the platform even though there is an approach to solving the subgroup membership problem in such a group mentioned above. We believe that this approach has only a theoretical advantage over the exhaustive search, and therefore it is not a threat to security of the cryptosystem.

In the interests of efficiency, we suggest to use a free metabelian group of a fairly small rank as the platform. The rank that we suggest is $r = n + m = 10$, where $n = 8$, $m = 2$.

Then, an important parameter is the number of "elementary" automorphisms in the factorization $\varphi = \tau_1 \cdots \tau_k$ of the key automorphism φ. One has to be careful here because, say, the automorphism group of a free group has exponential growth with respect to the (finite) generating set that consists of all Nielsen automorphisms. This implies, in particular, that the length of $\varphi(x_i)$ grows exponentially with k. However, in the free group of rank $r = 10$ the base of this exponent is so small (the higher the rank the smaller the base) that, according to our experiments, for a random product of 30 Nielsen automorphisms the average length of $\varphi(x_i)$ is under 30, which is quite reasonable.

Thus, we suggest $k = 30$, of which 90% should be random Nielsen automorphisms and 10% automorphisms of the type described in the Theorem in Section 3, i.e., $\alpha_{u,j} : x_j \to x_j[x_j, u, x_j]$, $x_i \to x_i$, $i \neq j$, for random elements $u \in [F_r, F_r]$. That means, when building the automorphism φ as a product of "elementary" automorphisms one factor at a time, Alice selects, at each step, a Nielsen automorphism with probability 90% and an automorphism of type $\alpha_{u,j}$ with probability 10%.

This provides for sufficiently large key space because there are approximately 100 Nielsen automorphisms in the free (or relatively free) group of rank 10, so products of 30 Nielsen automorphisms already give us approximately $100^{30} = 10^{60}$ choices for φ. On top of that, there are automorphisms $\alpha_{u,j}$ with arbitrary $u \in [F_r, F_r]$. The length of u has to be bounded by, say, 10, to keep the complexity of φ within reasonable limits. This leaves us with the pool of approximately 10^7 elements to choose u from.

With these parameters, the average total length of $\varphi(x_i)$, $i = 1, ..., 10$, is under 1000, according to our experiments.

Acknowledgement. We are grateful to Dennis Hofheinz for helpful comments and suggestions.

References

[1] I. Anshel, M. Anshel, D. Goldfeld, *An algebraic method for public-key cryptography*, Math. Res. Lett. **6** (1999), 287–291.

[2] S. Bachmuth, H. Y. Mochizuki, $Aut(F) \to Aut(F/F'')$ *is surjective for free group F of rank* ≥ 4, Trans. Amer. Math. Soc. **292** (1985), 81–101.

[3] G. Baumslag, R. Strebel, M. W. Thomson, *On the multiplicator of* $F/\gamma_c R$, J. Pure Appl. Algebra **16** (1980), 121–132.

[4] R. M. Bryant, C. K. Gupta, F. Levin, and H. Y. Mochizuki, *Non-tame automorphisms of free nilpotent groups*, Comm. Algebra, **18** (1990), 3619–3631.

[5] J. C. Cha, K. H. Ko, S. J. Lee, J. W. Han, J. H. Cheon, *An Efficient Implementation of Braid Groups*, ASIACRYPT 2001, Lecture Notes in Comput. Sci. **2248** (2001), 144–156.

[6] T. Coulbois, *Propriétés de Ribes-Zaleskii, topologie profinie, produit libre et généralisations*, Ph. D. Thesis, Université de Paris VII, 2000.

[7] B. Eick, D. Kahrobaei, *Polycyclic groups: A new platform for cryptology?*, preprint. http://arxiv.org/abs/math.GR/0411077

[8] D. Garber, S. Kaplan, M. Teicher, B. Tsaban, U. Vishne, *Probabilistic solutions of equations in the braid group*, Advances in Applied Mathematics **35** (2005), 323–334.

[9] M. Garzon, Y. Zalcstein, *The complexity of Grigorchuk groups with application to cryptography*, Theoret. Comput. Sci. **88** (1991), 83–98.

[10] L. Goubin, N. T. Courtois, *Cryptanalysis of the TTM cryptosystem*, Asiacrypt 2000. Lecture Notes in Comput. Sci. **1976**, pp. 44–57, Springer, 2000.

[11] D. Grigoriev, I. Ponomarenko, *On non-abelian homomorphic public-key cryptosystems*, preprint. http://arxiv.org/abs/cs.CR/0207079

[12] D. Grigoriev, I. Ponomarenko, *Homomorphic public-key cryptosystems over groups and rings*, preprint. http://arxiv.org/abs/cs.CR/0309010

[13] N. Gupta, *Free Group Rings*, Contemp. Math. **66**. Amer. Math. Soc., 1987.

[14] P. Hall, *On the finiteness of certain soluble groups*, Proc. London Math. Soc. **9** (1959), 565–622.

[15] D. Hofheinz, R. Steinwandt, *A practical attack on some braid group based cryptographic primitives*, in PKC 2003, Lecture Notes Comp. Sc. **2567**, 187–198. Springer, Berlin, 2002.

[16] K. H. Ko, S. J. Lee, J. H. Cheon, J. W. Han, J. Kang, C. Park, *New public-key cryptosystem using braid groups*, Advances in cryptology—CRYPTO 2000 (Santa Barbara, CA), 166–183, Lecture Notes Comp. Sc. **1880**, Springer, Berlin, 2000.

[17] R. C. Lyndon and P. E. Schupp, *Combinatorial Group Theory*, Ergebnisse der Mathematik, band 89, Springer 1977. Reprinted in the Springer Classics in Mathematics series, 2000.

[18] J. Milnor, *Growth of finitely generated solvable groups*, J. Differential Geometry **2** (1968), 447–449.

[19] T. Moh, *A public key system with signature and master key functions*, Comm. Algebra **27** (1999), 2207–2222.

[20] N. S. Romanovskii, *The occurrence problem for extensions of abelian by nilpotent groups*, Sib. Math. J. **21** (1980), 170–174.

[21] V. Shpilrain, *Automorphisms of F/R' groups*, Internat. J. Algebra and Comput. **1** (1991), 177–184.

[22] V. Shpilrain and A. Ushakov, *Thompson's group and public key cryptography*, Lecture Notes Comp. Sc. **3531** (2005), 151–164.

[23] V. Shpilrain and A. Ushakov, *The conjugacy search problem in public key cryptography: unnecessary and insufficient*, Applicable Algebra in Engineering, Communication and Computing, to appear.
http://eprint.iacr.org/2004/321/

[24] V. Shpilrain and G. Zapata, *Combinatorial group theory and public key cryptography*, Applicable Algebra in Engineering, Communication and Computing, to appear.
http://eprint.iacr.org/2004/242/

Department of Mathematics, The City College of New York, New York, NY 10031
e-mail address: shpil@groups.sci.ccny.cuny.edu
http://www.sci.ccny.cuny.edu/~shpil/

Department of Mathematics, CUNY Graduate Center, New York, NY 10016
e-mail address: nyzapata@verizon.net

Titles in This Series

418 **Lothar Gerritzen, Dorian Goldfeld, Martin Kreuzer, Gerhard Rosenberger, and Vladimir Shpilrain, Editors,** Algebraic methods in cryptography, 2006

417 **Vadim B. Kuznetsov and Siddhartha Sahi, Editors,** Jack, Hall-Littlewood and Macdonald polynomials, 2006

416 **Toshitake Kohno and Masanori Morishita, Editors,** Primes and Knots, 2006

415 **Gregory Berkolaiko, Robert Carlson, Stephen A. Fulling, and Peter Kuchment, Editors,** Quantum Graphs and Their Applications, 2006

414 **Deguang Han, Palle E. T. Jorgensen, and David Royal Larson, Editors,** Operator theory, operator algebras, and applications, 2006

413 **Georgia M. Benkart, Jens C. Jantzen, Zongzhu Lin, Daniel K. Nakano, and Brian J. Parshall, Editors,** Representations of algebraic groups, quantum groups and Lie algebras, 2006

412 **Nikolai Chernov, Yulia Karpeshina, Ian W. Knowles, Roger T. Lewis, and Rudi Weikard, Editors,** Recent advances in differential equations and mathematical physics, 2006

411 **J. Marshall Ash and Roger L. Jones, Editors,** Harmonic analysis: Calderón-Zygmund and beyond, 2006

410 **Abba Gumel, Carlos Castillo-Chavez, Ronald E. Mickens, and Dominic P. Clemence, Editors,** Mathematical studies on human disease dynamics: Emerging paradigms and challenges, 2006

409 **Juan Luis Vázquez, Xavier Cabré, and José Antonio Carrillo, Editors,** Recent trends in partial differential equations, 2006

408 **Habib Ammari and Hyeonbae Kang, Editors,** Inverse problems, multi-scale analysis and effective medium theory, 2006

407 **Alejandro Ádem, Jesús González, and Guillermo Pastor, Editors,** Recent developments in algebraic topology, 2006

406 **José A. de la Peña and Raymundo Bautista, Editors,** Trends in representation theory of algebras and related topics, 2006

405 **Andrew Markoe and Eric Todd Quinto, Editors,** Integral geometry and tomography, 2006

404 **Alexander Borichev, Håkan Hedenmalm, and Kehe Zhu, Editors,** Bergman spaces and related topics in complex analysis, 2006

403 **Tyler J. Jarvis, Takashi Kimura, and Arkady Vaintrob, Editors,** Gromov-Witten theory of spin curves and orbifolds, 2006

402 **Zvi Arad, Mariagrazia Bianchi, Wolfgang Herfort, Patrizia Longobardi, Mercede Maj, and Carlo Scoppola, Editors,** Ischia group theory 2004, 2006

401 **Katrin Becker, Melanie Becker, Aaron Bertram, Paul S. Green, and Benjamin McKay, Editors,** Snowbird lectures on string geometry, 2006

400 **Shiferaw Berhanu, Hua Chen, Jorge Hounie, Xiaojun Huang, Sheng-Li Tan, and Stephen S.-T. Yau, Editors,** Recent progress on some problems in several complex variables and partial differential equations, 2006

399 **Dominique Arlettaz and Kathryn Hess, Editors,** An Alpine anthology of homotopy theory, 2006

398 **Jay Jorgenson and Lynne Walling, Editors,** The ubiquitous heat kernel, 2006

397 **José M. Muñoz Porras, Sorin Popescu, and Rubí E. Rodríguez, Editors,** The geometry of Riemann surfaces and Abelian varieties, 2006

396 **Robert L. Devaney and Linda Keen, Editors,** Complex dynamics: Twenty-five years after the appearance of the Mandelbrot set, 2006

395 **Gary R. Jensen and Steven G. Krantz, Editors,** 150 Years of Mathematics at Washington University in St. Louis, 2006

TITLES IN THIS SERIES

394 **Rostislav Grigorchuk, Michael Mihalik, Mark Sapir, and Zoran Šuniḱ, Editors,** Topological and asymptotic aspects of group theory, 2006

393 **Alec L. Matheson, Michael I. Stessin, and Richard M. Timoney, Editors,** Recent advances in operator-related function theory, 2006

392 **Stephen Berman, Brian Parshall, Leonard Scott, and Weiqiang Wang, Editors,** Infinite-dimensional aspects of representation theory and applications, 2005

391 **Jürgen Fuchs, Jouko Mickelsson, Grigori Rozenblioum, Alexander Stolin, and Anders Westerberg, Editors,** Noncommutative geometry and representation theory in mathematical physics, 2005

390 **Sudhir Ghorpade, Hema Srinivasan, and Jugal Verma, Editors,** Commutative algebra and algebraic geometry, 2005

389 **James Eells, Etienne Ghys, Mikhail Lyubich, Jacob Palis, and José Seade, Editors,** Geometry and dynamics, 2005

388 **Ravi Vakil, Editor,** Snowbird lectures in algebraic geometry, 2005

387 **Michael Entov, Yehuda Pinchover, and Michah Sageev, Editors,** Geometry, spectral theory, groups, and dynamics, 2005

386 **Yasuyuki Kachi, S. B. Mulay, and Pavlos Tzermias, Editors,** Recent progress in arithmetic and algebraic geometry, 2005

385 **Sergiy Kolyada, Yuri Manin, and Thomas Ward, Editors,** Algebraic and topological dynamics, 2005

384 **B. Diarra, A. Escassut, A. K. Katsaras, and L. Narici, Editors,** Ultrametric functional analysis, 2005

383 **Z.-C. Shi, Z. Chen, T. Tang, and D. Yu, Editors,** Recent advances in adaptive computation, 2005

382 **Mark Agranovsky, Lavi Karp, and David Shoikhet, Editors,** Complex analysis and dynamical systems II, 2005

381 **David Evans, Jeffrey J. Holt, Chris Jones, Karen Klintworth, Brian Parshall, Olivier Pfister, and Harold N. Ward, Editors,** Coding theory and quantum computing, 2005

380 **Andreas Blass and Yi Zhang, Editors,** Logic and its applications, 2005

379 **Dominic P. Clemence and Guoqing Tang, Editors,** Mathematical studies in nonlinear wave propagation, 2005

378 **Alexandre V. Borovik, Editor,** Groups, languages, algorithms, 2005

377 **G. L. Litvinov and V. P. Maslov, Editors,** Idempotent mathematics and mathematical physics, 2005

376 **José A. de la Peña, Ernesto Vallejo, and Natig Atakishiyev, Editors,** Algebraic structures and their representations, 2005

375 **Joseph Lipman, Suresh Nayak, and Pramathanath Sastry,** Variance and duality for cousin complexes on formal schemes, 2005

374 **Alexander Barvinok, Matthias Beck, Christian Haase, Bruce Reznick, and Volkmar Welker, Editors,** Integer points in polyhedra—geometry, number theory, algebra, optimization, 2005

373 **O. Costin, M. D. Kruskal, and A. Macintyre, Editors,** Analyzable functions and applications, 2005

372 **José Burillo, Sean Cleary, Murray Elder, Jennifer Taback, and Enric Ventura, Editors,** Geometric methods in group theory, 2005

For a complete list of titles in this series, visit the
AMS Bookstore at **www.ams.org/bookstore/**.